AUSTRALIA'S MOST DEADLY and DANGEROUS BEASTS

AUSTRALIA'S
MOST DEADLY and DANGEROUS BEASTS

BRUCE THOMSON

Lothian BOOKS

Thomas C. Lothian Pty Ltd
132 Albert Road, South Melbourne, Victoria 3205
www.lothian.com.au

First published 2004
Reprinted 2005

Text copyright © Bruce Thomson 2004

All photographs © Bruce Thomson, with the exception of the following: Photographs © Gary Bell/Oceanwide Images on pp. ii–iii, 77, 79, 97, 99 and 101. Photograph © Chris Fallows/Oceanwide Images on p. 106–7. Photograph © ANT Photo Library 2004 on p. 105. Photograph © Lochman Transparencies on p. 73 (of water buffalo) and on p. 81.

All rights reserved. No part of this publication may be reproduced, stored in a retrieval system or transmitted in any form by any means without the prior permission of the copyright owner. Enquiries should be made to the publisher.

National Library of Australia
Cataloguing-in-Publication data:

Thomson, Bruce, 1956– .
Australia's most deadly and dangerous beasts.

ISBN 0 7344 0681 9.

1. Dangerous animals — Australia. 2. Dangerous animals — Australia — Identification. I. Title.

591.680994

Cover and text design by David Constable
Index by Russell Brooks
Printed in Singapore by Craft Print International Ltd

Disclaimer
The author has made every effort to ensure that the information in this book is accurate. However, the information and advice contained in this book are not intended as a substitute for consulting your doctor or health practitioner regarding any action that may affect your well-being. Individual readers must assume responsibility for their own actions, safety and health. Neither the author nor the publisher shall be liable or responsible for any loss, injury or damage allegedly arising from any information or advice contained in this book.

ACKNOWLEDGEMENTS

I wish to thank a number of people who contributed substantially to this book. I thank my wife and family for their support and, in particular, my son Jeffrey for his initial editing of the manuscript. Many of the snake photographs would not have been possible without the expert assistance of Terry Adams.

I would also like to thank Linda Thomas at the Institute for Molecular Bioscience, University of Queensland, for insights into the life of a cone shell and for allowing me to photograph captive specimens. Carol Dennis provided illustrations for the book and these are gratefully acknowledged. I also gratefully acknowledge the suggestions and advice provided by Martyn Robinson of the Australian Museum, Sydney.

Finally, I would like to acknowledge the enthusiasm and professionalism of the staff at Lothian Books, who have done such a wonderful job in bringing the project to fruition.

CONTENTS

Preface	ix
Spiders	1
Paralysis Ticks	21
Scorpions	25
Centipedes	27
Insects — Bees, Wasps, Ants and their Relatives	29
Other Garden Creatures	37
Dangerous Land Snakes	39
Dangerous Birds	67
Mammals	71
Cane Toads	75
Marine Jellyfish	77
Fish with Dangerous Spines	85
Other Spiny Marine Creatures	93
Cone Shells	95
Blue-ringed Octopus	97
Sea Snakes	99
Sharks	101
Crocodiles	109
Quick Reference — First Aid Summary	113
First Aid Techniques	116
Glossary	119
Index	123

PREFACE

AUSTRALIA IS RENOWNED AS HAVING THE MOST dangerous animals in the world but in fact many countries have wildlife that is at least as dangerous or, in some cases, more so than our own. However, increases in Australia's human population and the resultant increase in nature-based recreational pursuits have the potential to bring more of us into contact with possibly dangerous wildlife. More than ever, we need to educate ourselves about these dangers and understand how to avoid them — and what to do if our avoidance strategies fail.

Virtually all of Australia's potentially dangerous creatures would prefer to retreat rather than attack, and many close encounters involve a certain amount of unwitting assistance on our part, by acting inappropriately when dealing with dangerous wildlife or by simply not taking the necessary precautions to avoid contact when the possibility of an encounter exists. With improved knowledge, we can begin to understand more about the behaviour of our dangerous animals and avoid unnecessary risks.

The purpose of this book is to provide you with an insight into the characteristics and behaviour of a broad range of dangerous or harmful Australian creatures and the various ways in which you can minimise your chances of an unpleasant encounter. If avoidance has not worked, despite your best efforts, the book also explains, in simple terms, the symptoms that might be expected and some basic first aid measures that can be used.

The book will also help you determine if the bite or sting is serious enough to warrant hospital attention. Notwithstanding any advice given in this book, you should also use your personal judgement to determine if someone needs special medical attention. In so many cases, the perpetrator of the bite or sting is misidentified or not seen at all. Remember that it is always better to take too many precautions than not enough!

The Lions Mane Jellyfish, found in the temperate waters of Australia, is thought to be the world's largest jellyfish.

SPIDERS

A female Red-back Spider (*Latrodectus haseltii*). The spherical egg sacs are often seen before the spider.

SPIDERS ARE AN ANCIENT AND HIGHLY DIVERSE group, and there are well over 2000 known species in Australia, with many others awaiting discovery and description. They are found throughout the continent in virtually all environments, from below ground level to the tops of the tallest trees. There are even some species that can swim and are at home in freshwater environments. One such group is the genus

Dolomedes. These spiders can actually live for an impressive length of time underwater, breathing the air that remains trapped against their hairy bodies. They also capture underwater prey. This particular mode of aquatic life can be somewhat difficult, since all that trapped air makes them very buoyant. Getting under the water is usually an effort and, once there, they have to keep a firm hold of any available submerged objects, to prevent themselves popping back to the surface.

Since spiders are somewhat susceptible to water loss from their bodies when exposed to low humidity, the greatest number of Australian spider species is to be found in the moister coastal environments of the continent. There are, however, a considerable number of species in the arid zone and one of our largest 'bird-eating' or 'barking' spiders, *Selenocosmia stirlingi*, is found in this region. Many of these arid-land spiders have developed ways of minimising moisture loss, by living in deep burrows or foraging only at night when the daytime summer temperatures have abated.

The vast majority of our spiders are quite harmless to us and our pets, and they play a vital role in our environment as predators of insects and other small creatures. On the whole we have nothing to fear from spiders, although we should be aware of those species which are potentially dangerous and take the time to learn about them. If we do this, we can minimise the chances of being bitten.

IDENTIFICATION OF SPIDERS

Most of us would be familiar with the basic characteristics of spider anatomy, even if we can't assign the appropriate names to all the various parts! They have eight legs and a rather obvious abdomen, which is usually soft. At the rear of the abdomen are the two pairs of spinnerets from which web is extruded. The front of the body is known as the cephalothorax and it incorporates the head and the upper part of the body, or thorax. This part has a protective cover and is relatively hard. It can be shiny or covered in hairs. This arrangement of body parts is quite different to insects, which have a separate head and thorax, and thus have three main body sections.

At the front of the spider's cephalothorax are the mouthparts, which incorporate the palps — small finger-like appendages that pass food into the mouth — and the fangs. The fangs may work by a pincer action, where the tips move together, or they may be both inclined downwards and thus need to strike from above to secure their prey. On male spiders, one pair of palps has been modified to hold sperm and is used in mating. On some spiders these can look like an additional pair of legs.

With so many different species at large, the prospect of being able to identify them all is far beyond most of us. Even spider experts can have difficulties at times! We can, however, learn to recognise the potentially dangerous ones, since there are only a few in this category. We can also learn to distinguish the common, harmless species that we might find in our parks and gardens.

Probably the easiest way to recognise Australia's dangerous species is by their overall shape and behaviour. Almost everyone is familiar with the redback spider and its identification is relatively easy. Funnel-web spiders, however, are not easy to identify, since several other spiders in the same group look very similar. The golden rule is 'If in doubt, treat it as if it IS dangerous'.

THE MOST 'DANGEROUS' AUSTRALIAN SPIDER

Not many spiders in Australia can be considered highly dangerous. Of those that might fall into this category, the Sydney Funnel-web, *Atrax robustus*, is the undisputed winner. To be more specific, it's actually the male of this species that takes the honours. There are a number of other species of funnel-webs aside from the Sydney Funnel-web, but none of them is considered to be as dangerous, although it must be pointed out that the potential

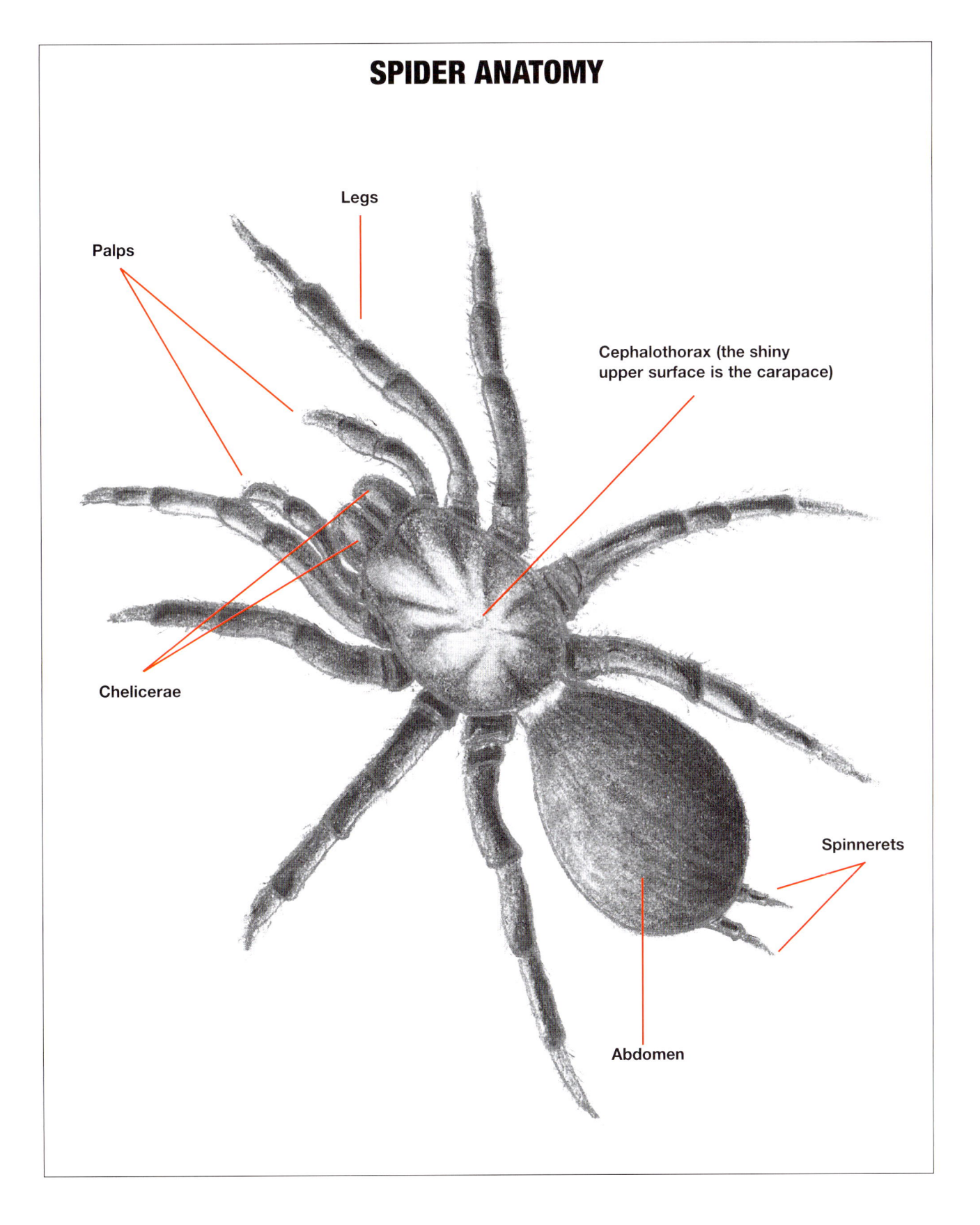

SPIDERS

SPIDER ANATOMY

- Legs
- Palps
- Cephalothorax (the shiny upper surface is the carapace)
- Chelicerae
- Spinnerets
- Abdomen

3

of some species to inflict fatal bites remains unknown. It is only in relatively recent times that species in the group have been concisely defined, and before that time funnel-web bites would not have been recorded against their appropriate species. The information that we do have indicates that from 1927 until 1980, when antivenom was developed, there were thirteen recorded deaths from funnel-web bites. Although it's difficult to establish with any certainty, there are an estimated 30–40 funnel-web spider bites each year in south-eastern Australia. Only 10 per cent of these cases require treatment, and there have been no deaths since the introduction of antivenom.

The Red-back, *Latrodectus hasseltii*, has been widely regarded as a dangerous spider and there were a number of deaths from bites of this species before the introduction of its specific antivenom in 1956. Since that time there have been no deaths, but the Red-back takes the prize for biting more people annually than any other venomous creature, including snakes, marine stingers and the like. Each year about 300 bites are recorded; however, the actual number is likely to be far higher! Of the bites that are recorded, less than 20 per cent of them cause significant symptoms.

HOW WE COMPARE WITH THE REST OF THE WORLD

Although many people regard the Sydney Funnel-web Spider as the world's most dangerous, the *Guinness Book of Records* bestows that honour on the Brazilian Wandering Spider or Banana Spider, *Phoneutria nigriventer*. These spiders are found in Central and South America, and they commonly shelter amongst banana leaves. They are large spiders with a body length of approximately 30 mm; when disturbed they aggressively defend themselves and will bite many times in rapid succession. Like the funnel-web, their venom is highly neurotoxic — it attacks the nervous system. Which is most dangerous? It's not easy to say.

Possibly the male Sydney Funnel-web has the more potent venom, but the Banana Spider may come into contact with people more frequently and account for more deaths each year.

The Red-back Spider has a geographic variant in the USA that is commonly called the Black Widow (*Latrodectus mactans*). It is possibly the most widely known dangerous spider in the world, but in fact various members of the 'red-back group' are found world-wide. In New Zealand their equivalents are called the Katipo and Night Stinger (*L. katipo* and *L. atritus*); both are endangered species. In Europe, their representative is known as the Malmignatte (*L. mactans tredecimguttatus*) and in South Africa they are known as Shoe-button Spiders. A few other species in the group have now been accidentally introduced to Australia and appear in some spider reference books. All of them have a very similar body form and venom with similar characteristics. None of the red-back or black widow spiders are as dangerous as the Sydney Funnel-web or the Brazilian Banana Spider.

POISONOUS SPIDERS

Funnel-webs, mouse spiders, trapdoors and their relatives (the Mygalomorphs)

The Mygalomorph suborder includes the funnel-webs, trapdoors, mouse spiders and Australia's largest species, the barking spiders. This suborder is one of the most primitive, with fossil specimens appearing as far back as the Triassic Period, 248–206 million years ago. They are primarily ground-dwelling spiders (although some may be found in the decaying trunks of trees) and construct silk-lined burrows which either have open entrances, or entrances that may be loosely closed with silk or sealed with carefully constructed lids.

Mygalomorphs are usually ambush hunters that are active at night, catching insects and other prey items near their burrow entrances. Many species are quite long-lived and have survived in captivity for up to twenty years.

Funnel-web spiders

There are two recognised genera, *Atrax* and *Hadronyche*, and approximately 40 different species in the 'funnel-web group' of spiders. They are found along the eastern Australian seaboard and along the southern coast as far as Adelaide, usually in moist habitats. The largest species is the Northern Tree Funnel-web (*Hadronyche formidabilis*) with a body length of up to 50 mm. As its common name implies, it lives in the hollow branches of trees in our wet forests and is one of only a couple of species that tend to be tree-dwellers. All the other funnel-webs follow the more typical lifestyle of the Mygalomorph group and live in the ground, where they construct silk-lined burrows that may be up to a metre long. Burrow entrances are typically placed in concealed locations amongst rocks, against tree stumps or under sheets of tin. The silk that lines the burrow is extended out from the entrance like a small sock and collapses to conceal the burrow entrance. After construction, the silk quickly becomes dirty and very effectively camouflages the entrance. Burrows are thus extremely difficult to see and are often only noticed when gardening efforts and cleaning-up

A funnel-web hole. The entrance has been opened up slightly for the photograph. It is normally concealed by the protruding portion of silk that lines the main tube. Leading from the burrow are radiating lines of thread that are used as trip lines to alert the concealed spider to the presence of passing prey.

AUSTRALIA'S MOST DEADLY AND DANGEROUS BEASTS

SPIDERS

activities tear the web from its anchor points and expose fresh silk. Interestingly enough, despite the spiders' common name, the burrow entrances don't look at all like funnels.

Lines of silk extend from the burrow entrance across the ground surface for distances of up to 15 cm and act as tripwires. At night, the spider sits just inside the burrow entrance and is alerted to the presence of potential prey when it moves against these trip lines. The spider then dashes out to ambush its prey and drags it back into the burrow for consumption.

When in their burrows, funnel-webs are normally shy and retiring and will avoid human contact if possible. If, however, their burrow is completely opened and the spider exposed, they will assume a defensive stance and will not hesitate to bite if a finger or other body part is placed within range.

Female spiders spend their whole life in or near their burrow and close encounters with them appear to be quite rare. The same cannot be said for the male spiders. Once males reach sexual maturity, they charge their palps with sperm, which is produced from organs on the underside of the abdomen and, ready to mate, they leave their burrows in search of female spiders. The males are able to detect a chemical scent or pheromone that is produced by the female, and can home in on a prospective mate from some distance. It is normally during these wanderings that males come into contact with people. This is rather unfortunate for us because in some species, such as the Sydney Funnel-web, the male's venom is known to be up to five times more potent than the females (although females produce more venom).

The venom of funnel-webs is also quite interesting because it is only potentially lethal to monkeys and man, since we lack a naturally occurring inhibitor that is present in most other vertebrates. So funnel-webs are not particularly dangerous to cats, dogs and other pets.

Male funnel-webs almost always move about at night when they are less obvious to predators such as birds, and when humidity levels are high enough to ensure that they don't lose too much body moisture. As the sky starts to lighten in the morning, they will move towards anything that can afford them cover for the day, and if clothing or shoes are left on the ground, these items will be used just as readily as any natural concealment option. As a result, many people are bitten when they put clothes or shoes on that have been left outside on the ground overnight. In fact, in their quest for females, the males will also clamber underneath loosely fitted doors to enter houses and may provide an unpleasant surprise when they are discovered in things left on the floor.

The other venue for close encounters is the swimming pool. Male spiders often fall into pools at night and will stay on the surface for a while, buoyed up by air trapped in their body hairs, but then sink to the bottom as the air is lost. Often, however, there is sufficient air still trapped to allow them to survive for some time, and so it should never be assumed that soggy spiders on the bottom of the pool are dead and safe to handle!

Identification

It can be difficult to identify the individual species of funnel-webs and it would be well beyond the scope of this book to attempt it. As a group, however, funnel-webs exhibit the following characteristics that, although not the most scientific, should suffice in most cases.

- The carapace is shiny black, never grey or brown. The abdomen is black, sometimes with a purplish sheen. Spiders may be brown immediately after they shed their skins, but this colour changes to black within a very short time. If placed in alcohol, these spiders will also go brown, and this may have given rise to the misconception that funnel-webs can be brown.

A female Sydney Funnel-web (*Atrax robustus*) showing the typical threat posture with forelegs raised. Venom droplets forming on the tips of the fangs are characteristic.

AUSTRALIA'S MOST DEADLY AND DANGEROUS BEASTS

- When initially provoked, a funnel-web spider moves back into a defensive pose with front legs raised and droplets of venom may be squeezed out onto the tips of the fangs, readily visible to the naked eye. Spiders that have been repeatedly harassed are unlikely to exhibit this feature because their venom stocks may have become temporarily exhausted.
- The two larger spinnerets at the rear of the abdomen are quite long: approximately 1 mm for a spider with a body length of 15 mm, or approximately 4 mm for a spider with a body length of 35 mm. The last segment of these spinnerets is also longer than wide. The paired spinnerets have between one and three segments.

Looking into the face of danger. The forward-looking eyes of this funnel-web reflect the silver light of the camera's flash. These spiders have large, heavily built fangs and a bite is a traumatic experience, quite aside from the effects of the venom.

- Male funnel-web spiders have a spur that is readily observed on the second pair of legs.

In the case of a bite, if any of these characteristics appear to match the spider involved, assume that it is a funnel-web. Apply the necessary first aid measures (see p. 9) and get the person to a hospital.

Avoiding funnel-webs

Once we understand something of funnel-webs' basic behaviour, it's not hard to develop an

SPIDERS

avoidance strategy. The most important points to remember:
- When venturing outside at night, always use a torch! A simple but often overlooked precaution.
- Never leave clothing or shoes lying on the ground overnight. If you have left items outside or in a shed, check them carefully before putting them on.
- Make sure doors and flyscreens fit securely, with no gaps.
- Check your swimming pool before swimming and fish any spiders out carefully, avoiding contact.

Bite symptoms

Most bites from these spiders are not serious enough to require the use of antivenom. However, the recommended precautionary approach is to treat the bite as potentially dangerous until medical staff determine otherwise. Appropriate first aid measures should be applied and the patient taken to hospital.

In cases of the Sydney Funnel-web, if significant envenomation has occurred the venom enters the circulation quite rapidly (within two minutes). The bite is painful and may remain so for many hours, or even days.

Within fifteen minutes of the bite, the patient may feel a tingling sensation around the mouth and may also start to salivate and produce tears. Abdominal pain, nausea and vomiting may occur and muscle spasms are common.

As the symptoms progress, shortness of breath may be experienced and blood pressure decreases then rapidly increases, followed by irregular heartbeat and, in some cases, cardiac arrest. Death may also be caused by fluid build-up in the lungs.

It sounds like the most horrific set of symptoms that one might ever be subjected to, but remember that since the introduction of antivenom no one has died from a funnel-web bite and symptoms can be greatly reduced by using appropriate first aid measures.

FIRST AID

- As with all bites, keep the patient as calm and relaxed as possible. The bite will be painful and being bitten is often a traumatic experience.
- Use a pressure immobilisation bandage, as described on p. 117.
- If possible, keep the spider for a positive identification, even if it has been severely squashed or dismembered. Put it in a bottle and add some alcohol (methylated spirits) if available.
- Seek urgent medical attention for the victim and take the spider, if it was caught or killed, for identification at the hospital.

A final note on funnel-webs

Recent research in the USA has claimed that individual funnel-webs — and possibly many other garden creatures — have distinct personality traits and that some individuals are naturally more aggressive and prone to behaving badly, whereas others are quiet and retiring! It was also discovered that spiders living in environments where there were few predators were bolder than those that lived in dangerous environments, where the risk of predation was higher. Does this mean that funnel-webs in urban environments are going to be more aggressive than those found in the bush? It's an interesting question!

AUSTRALIA'S MOST DEADLY AND DANGEROUS BEASTS

A shiny black carapace is characteristic of funnel-webs.

SPIDERS

Other Mygalomorphs

Other members of this suborder resemble funnel-webs in general body shape, being relatively heavy-set spiders. Often the male is smaller and more slender than the female. There are a considerable number of species in the group, the most noteworthy being the barking spiders or bird-eating spiders (*Selenocosmia*), which are Australia's largest species, the commonly encountered trapdoors (*Misgolas*, *Aganippe* and others) and the mouse spiders (*Missulena*). Mygalomorph spiders are found throughout mainland Australia and Tasmania.

Although the name 'bird-eating spider' might conjure up an image of a spider that builds a large web and catches birds in flight, this is far from the truth! These spiders are ground-dwelling and live in large, silk-lined burrows. Do they eat birds? It's quite possible, although such items wouldn't be a typical part of their diet. There are also other large spiders in the group, apart from *Selenocosmia*, that have been observed to eat small birds. Some years ago an incident was reported where a Double-barred Finch was taken by a very large trap-door spider. In this particular case, the finches were feeding on a lawn in south-eastern Queensland. Suddenly the birds flew up, but one remained fluttering on the lawn as if pinned to the ground. When the observer went to inspect, he found the bird close to death and as he slowly lifted the bird up to see what had happened, he found that a very large trapdoor had ambushed it and dragged it to its burrow entrance. Over the next few days the bird was slowly consumed.

A male Mouse Spider (*Missulena insignis*). Once males become sexually mature they are often found wandering in search of female spiders.

AUSTRALIA'S MOST DEADLY AND DANGEROUS BEASTS

A species of Barking Spider or Bird-eating Spider (*Selenocosmia stirlingi*) from central Australia. These are our largest spiders and live in burrows in the ground.

Reports of bites from this group of spiders are quite rare and symptoms described appear to be variable, possibly as a result of misidentification of the spiders responsible. For instance, the bite from the inland Bird-eating Spider (*Selenocosmia stirlingi*) has been described as being similar to a wasp sting, with localised pain and swelling that subsides after a few days. Then there was a bite attributed to the same species in Alice Springs where the patient exhibited quite severe symptoms: blinding headaches, extreme sensitivity to light, nausea and vomiting and tremors, all of which lasted for several days. Recent research has suggested that bites from the Eastern Mouse Spider (*Missulena bradleyi*) may also be very dangerous and it has been recommended that funnel-web antivenom be used in all cases where severe symptoms develop.

FIRST AID

Traditionally, first aid advice for these spiders has been focused on the treatment of local symptoms at the bite site to reduce pain and inflammation. A cold compress (see p. 116) applied over the bite will help. Pressure immobilisation bandages are normally not recommended, as they tend to keep the venom concentrated at the bite site and may increase the intensity of local pain and tissue damage.

In view of the more recent observations on the effects of some bites, however, it is recommended that close observation of the victim be maintained for some hours. If generalised symptoms begin to develop, such as headaches, dizziness or stomach pains, immediate hospital treatment should be sought.

If possible, keep the spider and if hospital treatment is needed, take it along with the victim, for identification by hospital staff.

SPIDERS

Red-back Spider

The Red-back Spider belongs to a genus (*Latrodectus*) which is now represented in Australia by a number of species. Some of these appear to be introductions from overseas and in fact there is some doubt about the status of our common and well-recognised species, the common Red-back Spider (*L. hasseltii*). Since it occurs in close association with urban environments and is largely absent in undisturbed bushland, it may also be an introduction from the early years of European colonisation. It has been recorded in all states and territories of Australia.

All species in this group share a similar body shape and all are venomous to some degree. The species that we need to be mostly concerned about, however, is the Red-back. This spider takes the award for having more bites recorded against its name than any other — approximately 300 each year.

The Red-back Spider (*Latrodectus hasseltii*) is common in urban environments.

Female red-backs normally construct a rather haphazard silk retreat on the underside of an object, in close proximity to the ground. In the most sheltered part of the retreat the egg sacs are placed, and leading down from the retreat to the ground are a series of silk lines. These lines are the primary entrapment mechanism of the structure and at night the spider will move down and sit just above the ground waiting for prey. When an insect stumbles into one of the lines, the spider is alerted and quickly moves in to wrap the prey in silk and bite

AUSTRALIA'S MOST DEADLY AND DANGEROUS BEASTS

The red stripe occurring on the Red-back's abdomen is quite variable and may be more orange on some spiders, as is the case with the female Red-back pictured here.

it. The prey is then elevated into the more protected part of the web and eaten.

Identification

Female red-backs vary in size, with a body length from about 6 mm to 15 mm. They normally have a distinct red stripe down the centre of the abdomen. This stripe is quite variable and may be more orange on some spiders or dark red and relatively indistinct on others. It may also have a white margin, particularly in younger individuals. The spherical, creamy coloured egg sacs are characteristic and often found before the spider is.

Male red-backs are much smaller than the females (only a few millimetres across) and can sometimes be found on the periphery of the female's web. They are typically black with lighter coloured stripes on their bodies and are quite harmless.

Avoiding red-backs

Be aware of the fact that red-backs are an ever-present part of our environment and make sure to check under chairs, pots and other objects before you pick them up. If this simple precaution is taken, most unpleasant encounters can be avoided. Remember that any sheltered, dark micro-environment may be used by red-backs: inside garden taps and pipes, under sheets of iron and so forth. As the song 'Red-back on the Toilet Seat' by Slim Newton pointed out many years ago, the underside of toilet seats in outdoor dunnies has always been a favourite location for red-backs. In the past, when these outdoor dunnies were commonplace, they were the major venue for bites and to add to the embarrassment, most bites were on the 'nether regions'! With the advent of modern indoor toilets, the overall incidence of red-back bites in Australia greatly decreased.

Bite symptoms

In most cases, red-back bites produce only localised symptoms of pain, inflammation and swelling. Occasionally symptoms may be more severe and require hospital attention. Children and infants are at far greater risk than adults, due to their smaller body size.

Symptoms usually develop slowly and may include an increase in the severity and extent of the pain, as well as nausea and vomiting and increased blood pressure. Localised sweating around the bite area is typical and diagnostic.

FIRST AID

A cold compress (see p. 116) over the bite site is recommended in order to reduce pain. If the symptoms become more severe, the victim should be hospitalised where antivenom may be administered. In some cases, allergic reactions may occur which lead to more intensified symptoms. In these cases, hospital treatment should be sought immediately. Deaths from red-back and other spider bites are more likely to be associated with allergic reactions than with non-allergic responses.

14

White-tailed Spiders, Black House Spiders and Wolf Spiders

These spiders are all common species found in urban and rural environments throughout Australia. The common names suggest that there are just a few different species, but in fact there are hundreds that may be grouped here. Even White-tailed Spiders are not a single species but rather a species complex of closely related forms.

Black House Spiders (*Badumna* species) are commonly found around window frames or in other situations around houses where a crevice or corner may provide refuge. In these sites they build quite conspicuous webs that have a distinctive funnel-shaped opening. This leads into the depths of the cavity or crevice where the spider hides. An insect blundering into the web will alert the spider which then runs out, collects its prey and retreats back into the safety of the web tunnel.

Black House Spider. The untidy webs of these species are commonly found around the windows of houses and can be distinguished by the presence of a central funnel that runs from the external part of the web, into a deep recess where the spider resides in relative safety.

White-tailed Spiders (*Lampona* species) don't use a web to capture prey, but actively hunt at night by wandering around, looking for likely victims. Other spiders, such as the Black House Spider, are often eaten by these vagrant hunters and it is not unusual to find white-tails in proximity to the latter's webs, tapping on the outer silk in an effort to entice the occupant out into the open. White-tails are often found in houses and although they move quite slowly when hunting, they can turn on quite a burst of speed to disappear into the first available refuge when disturbed. These spiders seem to be attracted to areas of high humidity and are often found in bathrooms.

The White-tailed Spider. Implicated in many cases of necrotic arachnidism, but recent studies show that it may have been falsely accused.

SPIDERS

Wolf Spiders (*Lycosa* species) may be reclusive, living in close proximity to their burrows, or they may be vagrant hunters like the White-tailed Spiders. A considerable number of species are found throughout Australia. Some of the burrow-dwellers construct lids for their burrows while others simply leave the entrances open. These spiders can move quickly and normally hunt insects on the ground surface that they locate visually. They often enter houses during the summer months and some species will defend themselves quite aggressively when threatened. Females usually tend their egg sacs in a burrow, but once the spiderlings hatch they are typically carried on their mother's back during her hunting sorties. People are often quite startled when they kill one of these spiders and then suddenly find that they have precipitated a mass stampede as baby spiders run everywhere!

Identification

Identification is best achieved with reference to the photographs in this book. Most of these species groups have distinctive body shapes and markings; however, colour patterns can be quite variable, even within a single species.

Bite symptoms

A range of symptoms have been described for bites from these species, from mild local pain which passes quite quickly, to ulceration and tissue damage around the bite site or, occasionally, quite severe general reactions including nausea. Possibly, the reason why bite symptoms appear to be so

Wolf Spiders are common ground-dwellers throughout Australia.

variable is that, in many cases, the spider is not actually seen and so misidentification of the culprit is probably a common occurrence. Thus the infamy of some species, such as the White-tailed Spider, is largely built on unreliable data. In fact many bites from the spiders discussed under this heading result in very mild reactions or none at all. In a recent study of 130 confirmed White-tailed Spider bites, none of the victims suffered any ulceration or necrotic tissue damage.

Even though there is some uncertainty as to which particular spiders cause this ulcerating effect, there is no doubt about the fact that it does occur and has been well documented by medical science. The process involves an ulceration or destruction of tissue around the bite site and is called 'necrotising arachnidism'. Why it occurs is still not fully understood. A number of theories, however, have been developed. In summary, they are:

- An effect caused by the spider's venom. This is known to be the case with spiders of the genus *Loxosceles*, and several species in America are particularly noted for causing massive tissue damage. A related species is now established around Adelaide and may be responsible for some bites that result in tissue necrosis. The venom of White-tailed Spiders, on the other hand, is not known to possess any tissue-destroying agents.
- An effect caused by the spider's digestive enzymes, which may have contaminated the fangs and been introduced into the wound made by the bite. This theory is not well supported by case studies and one would expect that if it was a possibility, then tissue damage would occur in most cases of spider bite since their digestive enzymes are all quite similar. In fact necrosis has not been recorded for spider bites from all species and so this theory seems a little improbable.
- An effect caused by a secondary infection carried by the spider's fangs and introduced into the bite. This would seem to be a plausible explanation, but in many cases no infectious organisms have been found when spiders' fangs have been examined. It is possible that the particular type of bacteria or fungi involved is difficult to isolate and identify, and so this theory may be supported by further research.
- An effect produced by the body's immune system. It has been suggested that the venom may cause an altered state in the affected human cells, such that our own body identifies them as defective and then destroys them, in much the same way as our bodies fight infection and replace worn-out cells. This failure of our immune system is known to cause a number of auto-immune diseases but has yet to be positively identified in the case of spider necrosis.
- As we get older, our bodies change and we may develop sensitivities to certain chemicals or other environmental stimuli. Allergic reactions are good examples of this. It is possible that in some people, spider venom may cause a similar type of localised reaction and result in a more severe effect than would normally be expected. Sometimes this effect may be exacerbated by external environmental factors that we have some sensitivity to; if these environmental factors occur regularly each year (as happens with pollen in spring) it may explain cases where bites break open or become inflamed on an annual basis for years after the bite.

FIRST AID

No specific first aid measures have been developed to treat bites from White-tailed, Wolf and Black House Spiders. A cold compress (see p. 116) may be applied over the bite site to control localised pain, but if more general symptoms develop, the victim should be taken to hospital. Every effort should be made to identify the spider responsible, and it should be bottled and taken to the hospital along with the victim if possible.

SPIDERS

HARMLESS SPIDERS
Orb-weavers

There is probably no worse sensation than walking into a large spider's web at night and then having to conduct a hurried search of one's clothing to locate the owner of the web. Usually this process is carried out with a degree of anxiety that ranges from controlled panic to outright hysteria! Perhaps it would ease our troubled senses if we knew that these spiders were relatively harmless — as is, in fact, the case. However, like the White-tailed Spider, they have been implicated (with very poor evidence) in some cases of necrotising arachnidism.

Large webs are usually made by the common Garden Orb-weaver (*Eriophora* species), Saint Andrew's Cross Spiders (*Argiope* species) or the Golden Orb-weavers (*Nephila* species). There are numerous other species that make smaller webs as well and orb-weavers can be found all over Australia.

Daddy Long-legs Spiders and associates

Another group of spiders that is commonly encountered is the Daddy Long-legs Spiders. These are probably familiar to all and are often found inside houses, where they construct very fine webs in the corners of rooms. When disturbed, these spiders will either drop to the ground and run away or sway furiously in their webs, the swaying action making their thin bodies and hair-like legs more difficult to see.

The Golden Orb-weaver Spider is common in parks and gardens. It is an inoffensive species and a reluctant biter.

The Daddy Long-legs Spider (*Pholcus phalangioides*). A common and harmless inhabitant of houses throughout the country.

While these are true spiders, there is another group of similar-looking animals that are also called Daddy Long-legs. These little creatures belong to an Order called Opiliones, whereas all true spiders belong to the Order Araneida. The Opilionids feed on decaying vegetative and animal material and have no fangs. The true Daddy Long-legs Spiders from the Order Araneida have all the same body parts as other spiders and belong to the Family Pholcidae, while Opilionids have eight legs but only one body part.

If we consider the true Daddy Long-legs Spiders for a moment, it is often said that they have the most deadly venom of all spiders but their fangs are too small to bite people. Is this true? No. There is no basis for this in fact. Firstly in relation to the toxicity of their venom, no researchers have ever collected Daddy Long-legs Spider venom and injected it into people to test the reaction. When you think about it, there would be no reason to do this. We don't tend to subject people to potential harm just to test out a theoretical proposition. On the second count of the fangs being too small to bite people, we know that spiders in the genus *Loxosceles* have a very similar fang structure and they can certainly bite people. So perhaps Daddy Long-legs Spiders have weaker muscles or some other feature that renders them less able to bite, but the truth is that we don't know for certain. We do know that there are no recorded bites from any of these spiders so they can most certainly be regarded as harmless.

Huntsman Spiders

Huntsman Spiders are usually tree dwellers; when found in houses they are usually on the upper walls or ceilings. These spiders are vagrant hunters and locate their prey by detecting air movements against the highly sensitive hairs on their legs. Vision may also play a part. In their natural habitat they often construct silk retreats under the bark of trees or they may pull leaves together to form a shelter. Females tend their eggs in these little shelters. When disturbed, Huntsman Spiders nearly always attempt to run away and very rarely if ever bite. Females guarding eggs will sometimes act aggressively to a perceived threat, but their apparent ferociousness is nearly always just a bluff.

Huntsman Spiders are active hunters. Their night-time sorties sometimes lead them into houses where they can be found in the morning, crouched motionless on walls or ceilings. Despite their appearance they are quite harmless.

PARALYSIS TICKS

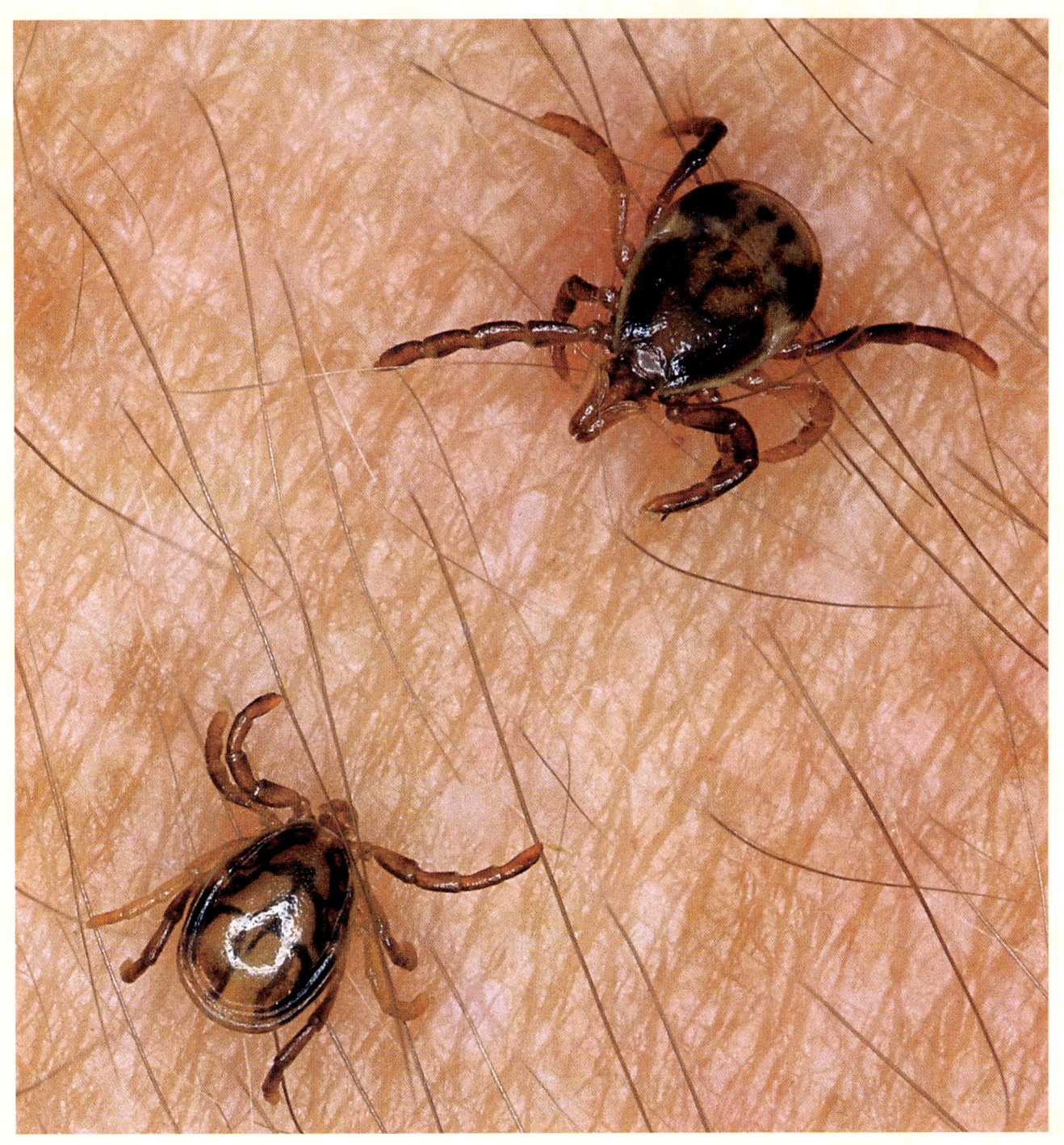

Male and female Paralysis Ticks on a host. The female at the top of the picture may be distinguished by its elongated, piercing mouth parts. Adult males do not feed on a host, but drop on board to search for and mate with the females.

21

AUSTRALIA'S MOST DEADLY AND DANGEROUS BEASTS

Ticks are placed in the same animal class as spiders (Arachnida), but are in a different Order (Acarina). In Australia there are approximately seventy species of ticks. Most of these are native species, with a few having been introduced with livestock. Almost all Australian mammals (even the platypus) have been recorded as hosts for various species of ticks and they are also found on snakes and lizards and a number of different types of birds. Of the seventy tick species, however, only four are thought to cause paralysis and of these, only one species, the Australian Paralysis Tick (*Ixodes holocyclus*), has been recorded as causing human deaths.

Distribution and life history

The Australian Paralysis Tick is found from the wet tropics around Cairns in northern Queensland, down the eastern seaboard to the Lakes Entrance area in eastern Victoria. It requires humid conditions and mild temperatures, and is usually most active in spring and summer throughout its distribution.

A second species, *Ixodes cornuatus*, also causes paralysis; it is found in Tasmania, Victoria and part of the alpine area of New South Wales. A number of other related species that cause problems for dogs, cats and other pets are found in many parts of Australia.

The Australian Paralysis Tick follows the same general life history pattern as other tick species, requiring three blood hosts to complete its life cycle. Female ticks lay between 2000 and 6000 eggs in leaf litter or low vegetation. These hatch as larvae and can be distinguished by having only three pairs of legs (as opposed to four pairs in subsequent life stages). The larvae climb up low vegetation and 'seek' a host. Once they attach to a passing host they feed from it for 4–6 days, then drop to the ground where they moult to become a nymph. The nymph attaches to another host and feeds for 4–7 days before once again dropping to the ground, where it in turn moults to become an adult male or female tick. The time taken for this process to occur is temperature-dependent, being faster in warm climates and slower in cold environments.

Once the ticks have reached adulthood, they again seek a host — the third and final one. This time the females attach to the host to acquire blood for egg production while the males move on to the host also as they are seeking females to mate with. Males rarely, if ever, feed directly from the host, but will pierce the female's body and feed from her. The female detaches after feeding and mating to lay eggs and complete its life cycle.

During all of its active life stages, when the paralysis tick is feeding it injects a toxin — which is carried in its saliva — into the host. The amount of toxin injected is dependent upon the size of the tick, and so normally the adult female ticks cause the most damage — they can cause severe paralysis and death in the host, if it is susceptible.

Bite susceptibility

Most native mammals act as hosts for the Australian Paralysis Tick and can carry a number of them with no apparent ill health. Some species, however, such as the Spectacled Flying Fox in the northern wet tropics of Queensland, have no natural immunity and die from infestation.

Pets, including cats and dogs, also have no immunity and will die from the effects of tick toxin. Livestock may be susceptible when young but most animals tend to build immunity with age.

People have no immunity to tick paralysis. The first officially recorded death from tick bite was documented in 1912 and at least twenty deaths occurred in the first half of last century. With the advent of modern medicine and tick antivenom, no contemporary deaths have been recorded. Those most at risk from tick paralysis are infants and young children, since their lower body mass can be more quickly affected by the volume of toxin being injected.

Ticks are also thought to act as vectors for the transmission of other diseases. Notable amongst these are rickettsial spotted fever (confirmed transmission), Q fever and Lyme disease (suspected transmissions).

PARALYSIS TICKS

Avoiding ticks

Ticks are normally found in vegetation, where they occur in close association with a range of host animals. When they are ready to move onto a host, they climb up grass stalks or into foliage where they wait with legs extended for a passing host, a behaviour called 'seeking'. When a suitable subject comes past, they detect its presence with an array of sensory receptors on the front part of the body, and then attach themselves as it goes past by either grabbing the host on contact or by falling or jumping. Once on the host, they move about to find a suitable site for attachment.

In areas where ticks are prevalent, they can be discouraged from climbing on board by the copious use of insect repellent on all parts of the body. You are also less likely to pick up ticks if you avoid brushing against grass and vegetation, including overhanging tree branches.

When you have been in a tick area, take the time to check your clothing and exposed parts of your body for ticks. They can be difficult to see — adult, unfed, female ticks are only 4–6 mm long while immature life stages are much smaller. You should also check your groin, armpits and hairlines, as ticks often bite at a 'barrier'.

Bite symptoms

Bites normally go unnoticed for a few days as the tick establishes itself at the bite site. As the tick feeds, it injects venom into the victim's body. Initially, the amount of venom injected is only small, but the venom increases over time and its effects become significant about day three or four. At this stage the tissue surrounding the point of attachment will have become hard and red, and swelling may make it appear as if the tick has burrowed in. Other localised symptoms — such as a red rash — may occur if the tick has introduced foreign organisms into the bite site (for example, rickettsial spotted fever) or more general symptoms such as enlargement and sensitivity of the lymph nodes. If the tick is removed at this stage, the swelling and sensitivity around the bite may persist for many weeks and a red inflammation may last for several months.

As venom injection proceeds, the victim may suffer paralysis of the muscles in the area around the bite and then a more general paralysis preceded by dizziness and lethargy.

In children, the symptoms to be aware of are lethargy and lack of coordination, loss of appetite and a tendency to sleep for longer than normal. As the poisoning progresses, the victim becomes progressively weaker and may have trouble seeing and speaking.

Some people are also highly allergic to ticks or tick by-products and can react quite severely without actually being bitten.

FIRST AID

Once the tick is located, the first action will be its removal. Although it has sometimes been recommended that the tick be first killed with methylated spirits or insecticide, this is not a preferred action since any irritation may cause the tick to disgorge its stomach contents and further venom into the victim. Disturbance to the tick should be minimised until the removal procedure is imminent. The objective of the procedure is to grasp just the biting mouthparts of the tick, without squeezing the body, and to ease it out. This can be accomplished with a fine pair of tweezers or a small pair of scissors.

In the case of tweezers, slide the tips in against the skin, grasp the tick by the head and then lift it out.

If using scissors, partially close the blades until a small 'V' is formed by the blades. Slide the V-shaped opening against the skin and, without letting the scissors close, move them in under the shoulders of the attached tick. Press down against the skin so that you contact the tick as close to the skin as possible, and slide forward to bring the tick out with steady pressure.

Sometimes the tick's mouthparts may be left in the wound and it is worth checking with a hand lens to make sure that the whole tick has been evicted.

Removal of the tick does not necessarily result in the immediate relief of symptoms and in fact the victim may become sicker before beginning to recover. In cases where a foreign pathogen has been injected by the tick, the victim may become ill a number of days after tick removal. If the victim starts to run a fever, breaks out in a rash, develops a headache or muscular pains, has sensitivity to light or develops a cough or sore throat, all of these symptoms may indicate a tick-borne infection and a visit to a GP or hospital is in order.

Antivenom is available for the treatment of severe tick bite cases after hospitalisation of the victim.

SCORPIONS

Some of our largest scorpions are found in northern and central Australia. This *Urodacus* species is a common inhabitant of the central Australian sand dunes.

Scorpions (Order Scorpiones) can be found in most parts of Australia, including Tasmania. There are twenty-nine different species recognised to date and they range in size from small, dappled creatures (about 1 cm long or sometimes less) that are often found in forest leaf litter, to the large brown or grey species that are found in tropical northern Australia. The latter species can be up to 12 cm long and they often dig a spiralling burrow with a characteristically domed ceiling and flat floor. They spend the day in these retreats and come out to hunt near the burrow entrance at night. They are often more active in wet or humid weather.

Our most venomous species tend to be the larger tropical ones but luckily we don't have anything that compares with some of the South American and African species that can deliver fatal stings. Australian species' stings are mild in comparison and produce local reactions around the site of the sting — usually pain, swelling and inflammation.

There is one record of a fatality in Australia attributed to a scorpion sting. Such cases appear to be exceedingly rare and there is almost no doubt that this person must have been allergic to the venom. As with any other type of bite or sting, if generalised symptoms start to develop — such as headaches, muscular pains, stomach cramps or difficulty in breathing — seek medical attention immediately.

FIRST AID

First aid is normally directed to the treatment of these localised symptoms with a cold compress (see p. 116) to alleviate the pain. The effects of the sting usually subside after a few hours, or in cases where the victim is particularly sensitised they may persist for a day or so.

CENTIPEDES

Centipedes are voracious nocturnal predators and a bite from a large specimen can be exceedingly painful. The jaws are under the head at the top of the picture. The two fleshy segmented appendages at the tail end are harmless.

CENTIPEDES (CLASS CHILOPODA) ARE ACTIVE nocturnal hunters of small insects and ground-dwelling creatures, including small skinks and frogs. They are common inhabitants of gardens and are usually found under rocks, logs and in leaf litter.

Some of our Australian species are quite small and may go unnoticed. Others, particularly in northern Australia, are quite large and may grow to a length of 15–20 cm. Centipedes' body structure is very primitive, consisting of a series of similar segments with a pair of legs extending from each one. The rear end of the body has two long, segmented appendages that are quite harmless. The head has a smooth, hardened upper dome with a pair of antennae at the front and a pair of claws on the underside that are modified into venom spurs. They have very poor eyesight and the antennae appear to be the main sensory organs involved in locating prey.

I can attest from personal experience that the bite from a large centipede is exceedingly painful, but luckily the effects are only local and the pain rapidly abates after a few hours. A cold compress (see p. 116) may be used to alleviate the pain.

INSECTS — BEES, WASPS, ANTS AND THEIR RELATIVES

Honey Bees (*Apis mellifera*) congregated in a shallow tree hollow prior to nest construction. Give bee nests a wide berth whenever you find them.

Honey Bees

THE INTRODUCED HONEY BEE *(Apis mellifera)* is found throughout Australia, where populations are 'domesticated' and managed by apiarists in artificial hives. Honey Bees have also escaped to form wild colonies and can now be found in suburban, rural and natural environments.

Honey Bees, like ants and many of the wasps, are colonial insects and work together to ensure the survival of their colonies. To do this effectively they have developed a division of labour that involves three castes — the workers, the drones and the queen.

Worker bees are reproductively underdeveloped females and, as their name implies, they do all the 'work'. That is, they make the hive cells, gather the nectar and pollen, defend the hive, feed the queen and carry out other maintenance duties. The queen is a fully developed female and her specialised task is the production of eggs. She also produces a number of chemical scents (pheromones) that regulate some of the hive activities. The drones are the males and although they might appear to do nothing, they do in fact have a task — they mate with the queens! Yes, you are excused for wondering if this social structure might in fact have been designed by a male!

Killer bees?

Bees used by the apiary industry originated from Europe and have been bred and managed for honey production for thousands of years. They are one of the least aggressive types of bees known. They do, however, have close relatives that evolved in other parts of the world and some of these are not so well behaved.

One particular sub-species is *Apis mellifera scutellata* or the 'Africanised Honey Bee'. These bees originated in East Africa and were imported to South America in 1956 as part of a breeding program to improve honey bee strains there. They escaped into the wild and have been steadily extending their distribution over the years. It is projected that their range will extend into much of the southern USA within the next decade or so. These bees are not found in Australia.

Although Africanised Honey Bees look almost exactly the same as their more mild-mannered cousins, they are much more of a danger to people — hence their rather sensationalist title of 'killer bees'. Some of the following traits are common to all bees, but are more pronounced in the African sub-species.

Africanised Honey Bees attack when they perceive a threat to their hive. As they attack, they release a strong chemical scent, or pheromone, that excites other bees in the hive to join the attack. An attack can be triggered by loud or high-pitched sounds, by strong odours or fragrances, by shiny jewellery or dark clothes. An attack may be initiated as far away as 30 metres from the hive and the bees will pursue for up to 500 metres. Once they become agitated, they may remain so for several days and attacks are even more likely during this period.

Avoiding bee stings

As has been pointed out, we do not have 'killer bees' in Australia, but if you are allergic to bee stings, it may be worthwhile taking note of the triggers that cause these bees to attack. To a far lesser extent, the same triggers may be relevant to our honey bees.

There are also other more obvious ways to avoid stings: wear shoes when walking on lawns where clover is flowering, don't wear sweet-smelling fragrances, don't brush into flowering trees where bees are gathering pollen and don't go near bee hives.

Sting symptoms

When a Honey Bee stings, its sting mechanism and associated venom sac break away and are left at the site of the sting. The bee then dies as a result of its attack. One or two bee stings are normally of minor concern and will produce only local pain, reddening

INSECTS — BEES, WASPS, ANTS AND THEIR RELATIVES

of the skin and some swelling.

A small percentage of people are allergic to bee stings and will be more seriously affected. Other people who are not particularly allergic may develop severe reactions if stung multiple times. In both cases, anaphylactic shock may result. This occurs because the venom causes histamine to be released from affected cells in the victim's body. Blood vessels open up and blood pressure then drops as a result.

This Paper Wasp (*Ropalidia* sp.) has made a nest in the cavity of a large tree. Some species will also build under sections of exfoliating tree bark.

Wasps

There are both native wasps and introduced species in Australia. Possibly the most dangerous is the introduced European Wasp (*Vespula germanica*), which is a communal species that builds its nest underground or in wall cavities. In recent years, this species has declined and disappeared from many parts of Australia where it had previously been recorded. It is distinctively yellow and black; however there are other native Australian species with similar colouring.

The native Masonry or Mud-dauber Wasps (*Sceliphron* species) are large yellow or orange and black wasps that build mud nests in sheltered positions, often in sheds and under house eaves. These wasps are solitary and quite inoffensive. They sometimes cause anxiety when they enter houses or inspect people at close quarters while outdoors. In both cases, they are probably looking for sites to build their nests or are looking for small spiders or caterpillars (depending on the wasp species) to stock their mud nests. These food items are placed in the nests for the wasp larvae.

FIRST AID

When removing the sting, always scrape it away with a fingernail or knife blade and never attempt to pull it out with tweezers, for instance, because the act of gripping the sting can compress the venom sac and cause more venom to be injected into the site of the sting.

If you have been stung multiple times or are dealing with a patient who has, be aware of the potential for this to happen. If any generalised symptoms start to develop, such as breathing difficulties or lapses of consciousness, immediate medical assistance must be sought.

People who are very allergic to bee stings may be carrying self-medication with them. Make sure that it is administered as per instructions and as quickly as possible after the stings.

Most Australian native bees are solitary and our native social bees are stingless, but the solitary species can sting many times as the sting is not barbed and does not detach from the bee. Fortunately, they rarely sting unless grasped or accidentally sat on.

AUSTRALIA'S MOST DEADLY AND DANGEROUS BEASTS

INSECTS — BEES, WASPS, ANTS AND THEIR RELATIVES

The other group of wasps with which most people are familiar are the native Paper Wasps (*Polistes* and *Ropalidia* species). These wasps are quite variable in colour: they may be dull brown and black with lighter banding or yellow and orange with black markings. These species are communal and build nests on the underside of a tree branch, in dense vegetation, on rock walls, under house eaves or in other similarly sheltered sites. Some species will build in hollow trees. Normally the paper-like nest is composed of numerous downward-facing cells and, like Honey Bees, the wasps tend these cells where the young are hatched and develop. These wasps will attack people who come too close, although our Australian species are relatively tolerant and can live quite harmoniously in suburban settings.

Avoiding stings

If you live in an area where European Wasps may be present, be aware that they are attracted to sweet substances. Many people have been stung on the mouth after these wasps have crawled into soft drink cans, so be careful!

Paper Wasps normally only attack when they perceive a threat to their nest. The best policy is to be aware of nests and to avoid them. If you need to walk near one, do so with slow and deliberate movements so as not to alarm the wasps.

Sting symptoms

Single wasp stings may cause local pain and inflammation. As with bees, a large number of wasp stings can produce more severe anaphylactic reactions. In these cases the victim may experience swelling of the tongue, cramps, nausea or diarrhoea and falling blood pressure.

Masonry Wasps (*Sceliphron* sp.) are solitary creatures that build mud nests. They are quite inoffensive and don't pose any threat when encountered in the garden, but individuals that are accidentally trapped inside houses should be treated with respect when being evicted.

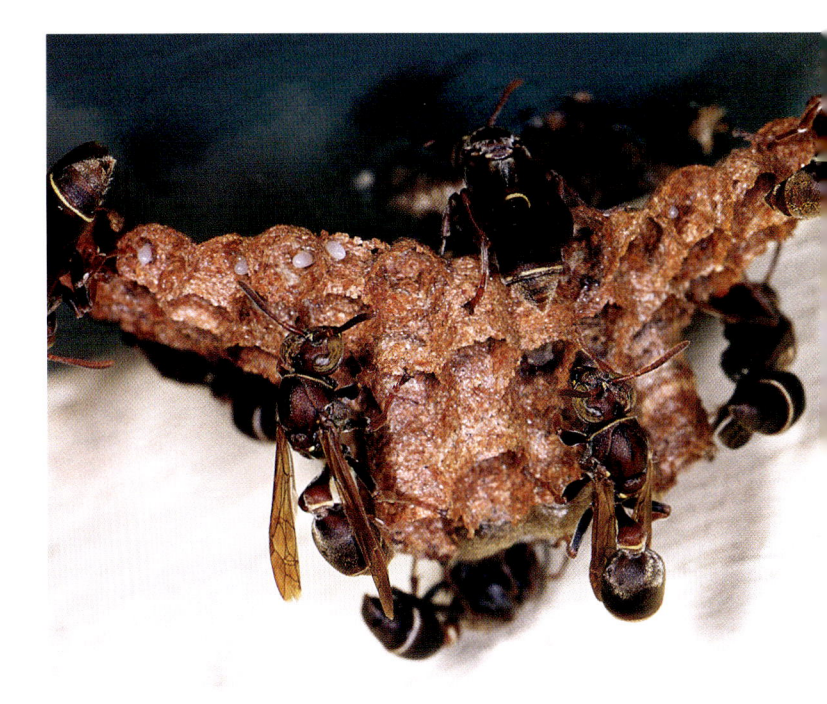

Paper Wasps (*Ropalidia* sp.) tending a small nest under the eaves of a house. They will attack people who get too close and stings can be quite painful.

If stung by a European Wasp the sting will typically be very painful and there will be local swelling. Anaphylactic shock may develop as described above, where the victim is stung multiple times. Single stings are less of a problem, although if stung on the mouth or inside the mouth, swelling may constrict breathing.

FIRST AID

Wasp stings normally produce only local pain and can be treated with a cold compress (see p. 116) to alleviate it. In cases where multiple stings are involved, be aware of the possibility of anaphylactic shock and seek hospital treatment immediately if any generalised symptoms develop.

The same precautions apply in the case of European Wasp stings. Victims who are stung in the mouth should be taken to hospital for treatment.

Ants

There are a number of native ant species that are known to inflict painful stings, and there are also two other species of concern that have been introduced from overseas.

Native ants

Native species include the jumping ant, bull-dog ant or inch ant (*Myrmecia* species), and the Greenhead

Green Tree Ants are common in northern Australia, where they build leafy nests in trees and shrubs. If you get too close, they may fire little jets of formic acid at you. Their bites are relatively mild.

Ant (*Rhytidoponera metallica*). The Greenhead Ant tends to be a solitary forager and is common in gardens and lawns. A bite from this ant can be very painful, producing a burning sensation at the bite site and sometimes a small blister. The effects are usually short-lived.

INSECTS — BEES, WASPS, ANTS AND THEIR RELATIVES

There are a number of species of jumping ants, most being closely associated with our native bushland, particularly eucalypt forests. In arid areas, species may occur in close association with stands of eucalypts or individual Bloodwood trees. All species build an underground nest with an elevated series of entrances on a small mound. The mounds are often covered with vegetative material such as gum nuts or sticks. These ants will aggressively defend their nests and move quickly towards an intruder by either running or by performing a series of rapid jumps. Once they have made contact with the victim, they will grip with their large, powerful jaws and then bring their abdomens underneath the body to sting numerous times with their stinger, which is located at the end of the abdomen. They can be extremely difficult to dislodge once they have a good hold with their jaws and, when scraped off, their heads will frequently remain attached to the skin.

Jumping Ants (*Myrmecia* sp.) are highly aggressive when their nests are threatened. These ants will grip firmly with their large pincers and then bring the tip of their abdomen up to sting numerous times.

Introduced fire ants

The imported Red Fire Ant (*Solenopsis invicta*), also known as the Southern Red Fire Ant, is of extreme concern not simply because of the injuries that these ants can inflict on humans, but because of the species' potentially devastating impact on natural environments. A native of South America, it is a highly aggressive species that often swarms and attacks en masse when nests or feeding areas are disturbed.

The Red Fire Ant was first identified in the Brisbane area in February 2001 but probably arrived in Australia through the Port of Brisbane some three to five years earlier. Nests can be large and appear as mounds of earth with no indication of the entrance holes (unlike the mounds of most native ant species). Mounds are often built up in winter to gather more thermal radiation from the sun. If the mound is dug up, it will reveal a hard, termite-like honeycombed structure that is quite distinctive and not found in any native ant species' nests.

Introduced Red Fire Ants (*Solenopsis invicta*) held in captivity at the Fire Ant Control Centre in Brisbane. Note the variation in size of the workers. This is one of the distinguishing characteristics of these ants.

A multi-million dollar eradication campaign appears to be winning the war against this ant, and it is hoped that it can be excluded from Australian dangerous wildlife references of the future. For now, however, it is included because it does have the potential to spread throughout much of Australia and its sting can produce quite severe reactions in people.

A second introduced species is the Tropical Fire Ant (*Solenopsis geminata*), which has become established in some coastal parts of the Northern Territory, but has not attracted much attention, even though it behaves in a very similar fashion to the Southern Red Fire Ant and is causing some concern in the Darwin area.

Ant bite symptoms

Native ants

Our Australian ant species can inflict quite painful stings and this is particularly the case with jumping ants. People who are allergic to these stings can suffer very severe reactions; one study has shown that quite a high percentage of our population (about 3–4 per cent) may be so affected. In most cases, however, the pain and swelling will subside after a day or two.

Introduced Red Fire Ants

Fire ants are considered by most experts to be much more of a problem than native ants. Like most ants, a single sting is manageable, but multiple stings can cause very severe reactions. Unfortunately with these ant species, multiple stings are common. Victims can also become hypersensitive to the venom and develop progressively more severe reactions over time if they are bitten on a number of occasions. When the ants attack, they grip with their pincers and bring the tail around under their body in very similar fashion to the jumping ants, and sting multiple times. Often they sting in an arc from the point of attachment of their jaws and so leave crescent-shaped sting patterns. Stings will produce a strong burning sensation and produce raised welts. These will develop into pustules and may remain for many days before subsiding into a red blotch. The potential for the victim to develop anaphylactic shock is considered to be much higher for this ant than for any other species. If this occurs, more generalised symptoms, such as difficulty in breathing and nausea, may become apparent and will need urgent medical attention.

FIRST AID

In all cases, a cold compress (p. 116) may be used to reduce swelling and alleviate pain.

In the case of fire ant stings, wash the stung area with soap and water, leaving the blisters intact. If scratched, they readily become infected and consequences can be quite serious.

Anaphylactic shock is a possibility when multiple stings are involved or when an allergic person suffers a single sting. These reactions can be life-threatening, so be aware of the possibility and if you are administering first aid to someone who has been stung, watch for the development of any general symptoms, that is, any that are not specifically related to the immediate area of the bite. If any such symptoms develop — such as laboured breathing, swelling in the throat or headaches — get the victim to hospital as quickly as possible. It's always better to take too many precautions than not enough!

OTHER GARDEN CREATURES

Assassin Bugs are found resting on the leaves of shrubs and trees and can deliver a painful sting if touched.

APART FROM SPIDERS, TICKS, BEES AND THEIR relatives, various other small invertebrate creatures that commonly occur in parks and gardens also have the potential to cause us considerable discomfort. The list is quite long — more so for people who may have allergies or be otherwise predisposed. Some of the most common ones are included here.

Assassin Bugs

A number of different species of assassin bugs are common garden-dwellers in most parts of Australia. They are hunters of other insects and are usually found on the leaves of plants, where they sit and wait for a suitable prey to wander by. These bugs have sharp, slender mouthparts and can deliver a very painful sting if hands or other exposed body parts get too close! The best treatment is a cold compress (see p. 116) to alleviate the pain and swelling.

Caterpillars

Many species of caterpillars have stinging or irritant hairs and can inflict considerable discomfort upon anyone who touches them. Usually the stings from caterpillars can be treated with a cold compress (see p. 116) to reduce pain and swelling. Those species with irritant hairs are a little more problematic because the hairs may break off and embed themselves in the skin, causing protracted discomfort. In some cases, the hairs can be removed by covering them with a rubber-based adhesive. Once dry, the adhesive is peeled off, with luck taking out the caterpillar's hairs at the same time. Other alternatives include sticky tape products, which can be placed across the hairs and pulled off to remove them from the skin. Care must be taken to ensure that the hairs are not pushed further into the skin when applying the material.

Beware of hairy caterpillars. Often the hairs are highly irritant and can cause skin rashes and itching.

DANGEROUS LAND SNAKES

Australia has 136 species of land snakes and about thirty of these are considered to be dangerous to people. Some of these 'dangerous' species are rather marginal in terms of the actual threat they pose, but they make sporadic appearances in references and books that deal with dangerous wildlife. However, it is unrealistic to expect that anyone, other than a dedicated snake enthusiast, would be able to confidently identify all of these species and know which ones to avoid. Thus the golden rule at all times is to leave snakes completely alone!

Delving into the topic of dangerous snakes also involves delving into a major area of Australian folklore and fantasy ... where a good story often takes precedence over the facts. So, enjoy the tales, but don't believe everything that you hear! A few good stories are debunked in the following pages.

IDENTIFICATION OF SNAKES

A glance through the photographs in this book should serve to emphasise the fact that Australian snakes have highly variable colour patterns, even within a single species. Identifications based on colour thus tend to be very unreliable and while this might be of little consequence when snakes are being identified for interest's sake, it's a more serious problem if a snake bite is involved. If the snake is dead, it may be taken to the hospital with the victim. If it's still rather healthy, leave it alone. One victim is enough! Over the past twenty years at least four deaths have directly resulted from people trying to kill snakes. It should also be remembered that snakes are native wildlife and are thus protected under State legislation, so killing them is illegal.

The development of modern medical procedures has now made precise field identification of snakes far less important. If the area around the bite site is left unwashed, an analysis of the venom taken from the skin around the puncture marks will provide doctors with the most reliable means of identifying the culprit. So apart from involving serious risk, the capture or killing of a snake involved in an attack is also no longer necessary in order to ensure appropriate hospital treatment.

Apart from difficulties in distinguishing various snake species, there is also the problem of distinguishing snakes from legless lizards! Although this might sound unlikely, a number of legless lizards look superficially like snakes, especially juvenile snakes. However, all legless lizards have an external ear opening that is reasonably easy to see. Snakes don't have these.

THE MOST 'DANGEROUS' AUSTRALIAN SNAKE

Which is the most dangerous Australian snake? A favourite topic that never ceases to draw in-depth discussion at barbecues, pubs and similar social gatherings and I would not wish to propose a definitive answer here that effectively quashes all further argument and associated folklore on the subject! Fortunately there is enough ambiguity in the question and enough unstated assumptions to allow a number of 'right' answers, and so there will be room for continued argument even after reading this book. Some of these assumptions need to be considered here, however, in order to provide at least a few insights.

What do we mean by 'dangerous'? If we were to consider the most dangerous type of animal in Africa, we would probably base our determination on the number of people killed over a set period by each of the animals under consideration. If we were to apply this definition to Australian snakes, the most dangerous are the brown snakes (genus *Pseudonaja*), having accounted for eleven deaths between 1981 and 1991. Second on the list would be the Eastern Mainland Tiger Snake (*Notechis scutatus*) with four deaths, and third would be the Coastal Taipan (*Oxyuranus scutellatus*) with two deaths in the same period.

Occasionally the terms 'most dangerous' and 'most venomous' get confused, the latter referring not to the number of deaths attributable to a species, but to an evaluation of the relative potency of snakes venoms. This is a very different issue, since a snake species might have a very potent venom but occur in a sparsely populated region and be disinclined to bite when encountered; it would therefore rate highly on this ranking but very low on the 'dangerous' ranking. Such is the case with the Inland Taipan (*Oxyuranus microlepidotus*) which is recognised as the world's most venomous snake. In fact Australian snakes dominate the world listings in the Most Venomous category. Their Australian ranking is as follows:

1. Inland Taipan (*Oxyuranus microlepidotus*)
2. Eastern Brown Snake (*Pseudonaja textilis*)
3. Coastal Taipan (*Oxyuranus scutellatus*)
4. Eastern Mainland Tiger Snake (*Notechis scutatus*)
5. Black Tiger Snake (*Notechis ater*)

Because it is somewhat unethical to inject people with measured amounts of venom to determine

DANGEROUS LAND SNAKES

levels of toxicity, mice are normally used for this purpose. The above rankings are therefore based on mice reactions rather than human. The assumption is that humans will react in much the same way but, as we have seen with funnel-web spiders, this is not necessarily the case. The reality is that humans may find some of these poisons to be relatively more or less toxic than mice do, and so the rankings may be considered as indicative but not necessarily exact.

Finally, there is another way of defining 'dangerous' and that is to consider the species in terms of their potential ability to inflict a lethal bite. In this case, we need to consider a number of attributes, including the proximity of the species' distribution to human population centres, the snake's tendency to bite when encountered, the toxicity of the venom and also the amount that is normally injected. The volume of venom that a snake produces can normally be correlated with its size, and so the larger the snake, the more venom it should have.

If we were to consider these criteria, then the number one dangerous snake would almost certainly be the Coastal Taipan. Second on the list would be the Eastern Brown Snake and third would be the Eastern Mainland Tiger Snake.

HOW WE COMPARE WITH THE REST OF THE WORLD

Although Australia may have some of the most venomous snakes in the world, we certainly don't rank highly in terms of the number of deaths due to snakebite. It is estimated that there may be about 3000 cases of snakebite in Australia each year. Of

The Coastal Taipan (*Oxyuranus scutellatus*), possibly the world's most dangerous snake. The pale face is a common feature of these snakes but at least one other relatively harmless species has the same coloration.

these, only about 300 require treatment with antivenom and, on average, we have one to three deaths each year. This is an amazingly small number, since there are an estimated one to two million snakebites in the world each year, resulting in 30,000 to 60,000 deaths!

The country that tops the list for snake-related deaths is Sri Lanka, where between 800 and 1000 people die each year. The main snake species responsible for the tally are the Indian Cobra (*Naja naja*), Russells Viper (*Vipera russelli*) and the Indian Krait (*Bungarus caeruleus*).

If we consider the relative toxicity of snake venoms and compare Australian species with those in other parts of the world, then we can claim some fame! The listing for the top ten varies somewhat between authorities, but looks something like this:

1. Inland Taipan (*Oxyuranus microlepidotus*) (Australia)
2. Eastern Brown Snake (*Pseudonaja textilis*) (Australia)
3. Coastal Taipan (*Oxyuranus scutellatus*) (Australia)
4. Eastern Mainland Tiger Snake (*Notechis scutatus*) (Australia)
5. Black Tiger Snake (*Notechis ater*) (Australia)
6. Beaked Sea Snake (*Enhydrina schistosa*) (Australia)
7. Western Mainland Tiger Snake (*Notechis scutatus occidentalis*) (Australia)
8. Black Mamba (*Dendroaspis polylepis*) (East Africa)
9. Black Tiger Snake (*Notechis ater serventyi*) (Chappell Island, Australia)
10. Southern Death Adder (*Acanthophis antarcticus*) (Australia)

The Indian Cobra (*Naja naja*) comes in at number 13, the King Cobra (*Ophiophagus hannah*) at number 18 and the Eastern Diamondback Rattlesnake (*Crotalus adamanteus*) at 24.

There is no definitive data on which snakes might top the list of the World's Potentially Most Dangerous, but no doubt the Coastal Taipan and the Eastern Brown Snake would rank very highly, possibly coming first and second.

AVOIDING UNPLEASANT ENCOUNTERS

Nearly all species of snakes are active hunters of small reptiles, amphibians, birds and mammals, with dietary preferences varying between individual species and also depending on the size of the snake itself. Suburban and rural gardens, sheds and bird aviaries provide ideal environments for their prey, and so it's not surprising that snakes are also found in these places. Recent research on tiger snakes has found that they are particularly adept at living in urban environments whilst remaining out of sight and undetected by humans.

Some of our most dangerous species, the taipans and brown snakes, are small mammal hunters and are attracted to bird aviaries and chook runs where grain and pellet foods support healthy populations of mice and rats. Moisture is also a critical requirement for snakes, especially in dry regions or during the hot summer months, and so our relatively cool, watered gardens and lawns are also important attractants. A further requirement for snakes is shelter and dense vegetation; stacks of wood or stones, iron and other similar objects are all suitable sites for snakes to escape from predators or the midday heat.

You can take action to reduce the attractiveness of outdoor areas for snakes by removing some of these magnets:

- ensure that all bird food (grain, pellets, etc.) is kept in sealed containers
- control populations of rats and mice by poisoning or trapping
- remove dense grass and other vegetation, and clean up woodheaps and other forms of cover that snakes might use.

There are a number of other commonsense precautions that you can take at all times:

- Always ensure that you can see where you are walking when in the bush. Don't step over a log without being able to see the ground where your foot is about to fall, and don't run through long grass.
- At night, always make sure you use a torch and, once again, watch where you are walking. Many of our dangerous snakes are active during the day, but some will hunt on warm nights. Included in this group are the black snakes, tiger snakes, death adders and, occasionally, some of the brown snake species.
- If you live in a rural area and expect to be walking in areas where snakes might be present, be sure to wear good sturdy shoes and long trousers. A constantly recurring theme that appears in

many contemporary accounts of snakebite fatality is the lack of recognition by the victim that he or she has been bitten. In most of these cases, the victim starts to feel ill and then makes the comment to bystanders, 'I wonder if that snake that I saw two hours ago might have bitten me?' In these situations, the venom has had time to spread, and the onset of symptoms may indicate damage that may be difficult or impossible to reverse. Be aware that if you are engaged in any physical activity, you may not necessarily feel or see a snake bite. If you think that you may have had a close encounter with a snake, stop and take the time to examine legs or other body parts that may have been exposed. Early recognition that you have been bitten means you can start applying first aid, and this may save your life.

In describing the behaviour of snakes, terms such as 'aggressive' or 'angry' are often used. We need to remember that these are essentially human words to describe human behaviour. In people, such behaviour may be the culmination of quite complex thought processes and we tend to assume a degree of intent accompanies the behaviour. In people, we also know that aggressive or angry behaviour is optional and that, because we have flexibility and control over our thought processes (at least to some degree), there may be alternative responses available. The situation is entirely different for snakes and other animals with less developed mental capacity.

Snakes have a relatively poorly developed brain and are simply not capable of thought processes as we know them. Their behaviour is almost completely innate — passed down through their genetic coding from previous generations. Their behaviour is therefore quite inflexible and snakes don't have the option of being able to judge the circumstances of an unfolding event, determine if it appears to be a threat and then evaluate the various options for escape! They simply perceive an event and then spontaneously react to it. The reaction might be to move away as rapidly as possible or it may be to go into what we might call a 'defensive' posture; if the threat is too close, it might involve an initial attack, which we interpret as 'aggressive', and then an escape. In all cases, however, we are the only party involved in the interaction with the ability to think and possibly to defuse the situation by modifying or controlling our behaviour.

So if you suddenly find yourself in close confrontation with a snake, what should you do to defuse the situation? The first thing is to remain perfectly still. Snakes are particularly alert to movement and if you stay still for a while, it seems that they start to have trouble distinguishing you from the rest of their surroundings and they move away. If you can't stay still, then back slowly away with minimal movements. You would be surprised how often those simple measures can get you out of danger. You would also be surprised to learn that many people foolishly attempt to do the opposite — they either move quickly towards the snake or even try to pick it up! People have died in Australia doing this, so don't add to the tally!

'NESTS' OF SNAKES?

A common belief abounds that if you find a small snake, then there must be a 'nest' nearby, and you can expect to see more and also the mother. There is no real basis for this belief. Many of our dangerous snakes are oviparous — they lay eggs — and once the eggs have been deposited the female snake leaves the area. In some cases, for instance death adders, where the species does not have a large home range and is not particularly mobile, the parent might remain in the general vicinity. This is not a maternal behaviour. Once the young are born, they disperse from the site and do not return or congregate nearby.

The only variation in this pattern occurs with pythons, where the female will remain with the eggs and assume a maternal function. Upon hatching, however, all progeny and the adult will disperse.

So, at least in Australia when we are talking about dangerous snakes, there is no such thing as a 'snake nest'! There is, however, some tenuous ground for referring to a python with eggs as 'nesting'.

INTER-BREEDING

There is also another interesting belief that dangerous snakes can mate with members of another species and, in so doing, produce more virulent offspring (presumably with venom properties that combine the worst of both parent species!). This is not the case: snakes are like all other animals in this respect. They do not inter-breed in the wild.

The marked variation in colours and patterns within a single species may lead some people to believe that they are witnessing an act of crossbreeding. You can be assured that if you see two snakes of very different colours entwined in breeding, they are of the exact same species.

In some rare cases, snakes in captivity may be made to crossbreed with other very closely related species, but the young are often less vigorous and fail to survive, or they are infertile and can't continue the lineage.

SNAKEBITE

Specific snakebite symptoms are described for each of the species in their individual accounts below. However, a few general observations may be covered here.

General symptoms

Available statistics suggest that the majority of snakebites in Australia involve 'dry bites', where no venom is injected. A percentage, however, does involve venom and in these cases the symptoms of the victim will vary, depending upon the species of snake and the location and severity of the bite.

Snake venoms are composed of various toxic substances (mainly proteins) that produce a range of effects in the victim. Venoms may attack the nervous system (this is called neurotoxic) and cause paralysis, or may cause blood clotting (pro-coagulant) or the thinning of blood (anti-coagulant). They may also cause tissue damage (cytotoxic) or muscle damage (myotoxic).

Most snake venoms are dominated by one or two substance groups, but also carry a lesser mix of other substances. Often closely related species have similar venom properties and so the brown snake group (genus *Pseudonaja*) tends to cause neurotoxic and pro-coagulant reactions whereas the black snake group (genus *Pseudechis*) tends to produce myotoxic, cytotoxic and pro-coagulant effects.

FIRST AID

In all cases of snake bite it is very important to ensure that the bite site is NOT WASHED. Most hospitals have venom identification kits and can quite reliably determine the species of snake from the residual venom that is left on the skin.

This is extremely important because an accurate identification allows the use of a specific antivenom, which is more effective in treating the victim and causes fewer side effects than a polyvalent or non-specific antivenom.

If the bite is on a limb (and most are), a pressure immobilisation bandage should be applied as quickly as possible. A full description of this technique is provided on p. 117.

The fact of being bitten by a snake is often quite traumatic for the victim and every effort should be made to calm the person. A reminder that these days almost no one dies from snakebite when medical attention is sought, might help. Discourage physical activity and get the victim to relax as much as possible but don't encourage sleep. The person needs to be awake so that they can communicate and you can gauge their condition.

DO NOT incise the bite area in an attempt to get it to bleed. This is a potentially dangerous practice and it does very little to remove venom. DO NOT attempt to suck the venom out.

DO NOT apply a tourniquet in the form of a strap or band on the limb above the bite site. This may block the supply of blood to the limb and cause major complications.

DANGEROUS LAND SNAKES

TAIPANS

Coastal Taipan

The Coastal Taipan (*Oxyuranus scutellatus*) is one of the most dangerous of our Australian snakes. It is found along the east coast from the Brisbane area to the top of Cape York in Queensland, and across the northern tropics of the Northern Territory, often close to settled areas.

These snakes are highly active daytime hunters with very good eyesight. They feed primarily on rodents and other small mammals, which they locate initially by smell and then, in the closing moments of the hunt, they track the prey visually. Coastal Taipans are seldom encountered, not because they are all that rare, but because their keen senses are able to detect the presence of people and other large animals at a considerable distance. By the time we stumble across the spot where the snake was, it has long gone!

Coastal Taipans are sometimes found sunning themselves on cold mornings in northern Australia, but they are extremely wary and will often move away well before they are approached closely enough to be seen.

Occasionally close encounters do occur and they are probably as much of a shock to the snake as they are to the person involved. Under these circumstances the snake may show little hesitation in striking and this species has a tendency to bite with little provocation.

Coastal Taipans have relatively long fangs and inject considerable quantities of venom (an average Coastal Taipan venom yield is around 120 mg; other Australian species yield less than 50 mg). Taipans tend to bite and release when they attack, and so a person may be bitten multiple times — the bites are delivered so rapidly that the victim may not be aware of all of them at the time of the encounter.

Coastal Taipans are considered to be the most potentially dangerous snake in the world and only one person survived a bite from this species before the development of antivenom. Even now, with modern treatments, bites need to be treated very quickly to maximise the chances of survival.

The largest Coastal Taipan ever officially measured was one that was raised in captivity from a 25 cm hatchling. Its owner, Joe Sambono, named it Terrence and when it died, some seven and a half years later, it was 2.9 metres in length. The specimen is now kept at the Queensland Museum while a cast of it was prepared and is on display at the Cooktown Natural History Museum.

VENOM PROPERTIES

The venom is highly neurotoxic and causes paralysis in victims, leading to a cessation of breathing and death. Other components of the venom are pro-coagulant (causing the blood to thicken and clot) and myotoxic (causing serious skeletal muscle damage).

Inland Taipan

Like the previous species, the Inland Taipan (*Oxyuranus microlepidotus*) is an active daytime hunter. It is found on black cracking clay soils in the Diamantina River drainage system in south-western Queensland. It is a specialist hunter of small mammals, particularly the Long-haired Rat (*Rattus villosissimus*). These rats can breed up to large numbers during good seasons. The snake's venom is the most toxic of any land snake venom known; this may be an important adaptation for a snake that specialises in catching rats and has to kill quickly to avoid being injured from their bites and scratches.

These snakes are less inclined to bite when encountered than the Coastal Taipan and their distribution in a relatively sparsely settled part of Australia means that encounters are rare.

The Inland Taipan (*Oxyuranus microlepidotus*) has the most toxic venom of any snake but is a relatively placid species and will seek shelter rather than a fight when encountered.

DANGEROUS LAND SNAKES

BROWN SNAKES

Eastern Brown Snake

This is one of the most common snakes in Australia and the species that has claimed the most human fatalities over the period in which records have been kept. The Eastern Brown Snake (*Pseudonaja textilis*) is distributed along the eastern coast of Australia, extending inland to western New South Wales and western Queensland. The southern part of its range covers most of Victoria and the south-eastern parts of South Australia. An isolated population occurs in central Australia around the MacDonnell Ranges.

Eastern Brown Snakes are primarily daytime hunters although they may occasionally be found out and about on warm summer nights. They will take a variety of prey, but feed predominantly on small mammals, birds and reptiles. They are primarily ground-dwellers but will climb small shrubs and trees when hunting and may be encountered on the rafters of sheds and in tree hollows. Prey is located by scent and vision and the snake typically strikes and maintains a hold, then quickly wraps its forebody around the prey to subdue any initial struggling.

Eastern Brown Snakes (*Pseudonaja textilis*) are agile daytime hunters and should be treated with a great deal of respect. They can be lightning fast and are just as likely to move towards a potential threat as they are to retreat.

This Eastern Brown Snake shows the typical 'S'-shaped defensive stance which is characteristic of the brown snake genus.

AUSTRALIA'S MOST DEADLY AND DANGEROUS BEASTS

This adult Eastern Brown Snake still has quite obvious vestiges of the banding that it had as a juvenile.

Another example of colour variability: this Eastern Brown Snake has a dark neck band and relatively plain brown body.

Eastern Browns are unpredictable snakes in terms of behaviour: they can be relatively placid or very active and fast, ready to strike at a moment's notice. A defensive pose with the forebody raised off the ground and forming a figure 'S' is typical of this species, as is the pale yellow underside with orange or light brown blotches. When striking, the snake may bite and release or it may adopt the approach it uses on food items, holding on and chewing.

Eastern Browns are egg layers and produce clutches of 10–35 soft-shelled eggs. Young snakes have dark head markings and look very much like some of the common species of legless lizards. In some cases the body may be striped with dark rings. Adults are highly variable in colour and sometimes carry the vestiges of their juvenile markings, although most assume a uniform dusky brown colour on maturity.

Juvenile Eastern Brown Snake. The juveniles of most brown snake species have distinctive markings. There is also a considerable degree of variability, with some individuals banded in black. This particular snake has a uniform pale brown body.

VENOM PROPERTIES

Bites from Eastern Browns are particularly dangerous because the initial symptoms resemble a typical shock response, and are thus not recognised as the result of the envenomation.

The more serious symptoms are slow to develop, but when they do start to manifest they progress quickly and the victim may die quite suddenly and unexpectedly. The poison is strongly neurotoxic and pro-coagulant.

Treat all brown snake bites as quickly as possible, with full pressure immobilisation bandages (see p. 117) and seek hospital attention immediately.

DANGEROUS LAND SNAKES

Western Brown Snake

The Western Brown Snake (*Pseudonaja nuchalis*) is generally found throughout Western Australia, the Northern Territory, South Australia and the western parts of Queensland and New South Wales. Like other brown snakes, this species is an egg layer; it produces 13–22 eggs in each clutch.

Adult snakes have several colour forms, which range from a uniform brown through reddish-brown with a black head, to individuals that are broadly banded in shades of brown. A helpful identification character that separates this species from the Eastern Brown Snake is the blackish colour of the inside of the mouth — the buccal cavity. In the Eastern Brown, the inside of the mouth is pinkish.

Western Browns are very active hunters during the day but may also be found on warm summer nights. They are not a particularly aggressive species but when closely approached will raise their forebody in an 'S'-shaped curve and strike repeatedly.

VENOM PROPERTIES

At least four deaths attributed to this species have been recorded in Western Australia since 1980. The venom is not particularly neurotoxic but rather procoagulant (causing the blood to thicken and clot) and myotoxic (causing damage to muscle tissue, in this case, often damage to the heart).

Western Brown Snakes (*Pseudonaja nuchalis*) exhibit considerable colour variation and above are three examples: (from top) a coppery individual with a black head, a plain brown variation and a banded form. This graphically illustrates the difficulties inherent in relying on colour as the sole means of identification.

AUSTRALIA'S MOST DEADLY AND DANGEROUS BEASTS

Speckled Brown Snake

The distribution of the Speckled Brown Snake (*Pseudonaja guttata*) is restricted to the black soil plains of the Barkly Tablelands in the Northern Territory and the Diamantina drainage basin in western Queensland.

Speckled Browns are most active during the day and appear to be predominantly reptile hunters, although doubtless other prey is taken when an opportunity presents itself. The black soil country where this species lives is characterised by Mitchell Grass (*Astrebla* species) grasslands on soil that is deeply cracked and fissured. The cracks and fissures provide the snakes with shelter.

Due to this species' distribution in a relatively sparsely settled part of the continent, encounters with people are not common and, as a result, bites are quite rare.

Speckled Brown Snakes (*Pseudonaja guttata*) are another species restricted to the black soil plains of western Queensland and north-eastern Northern Territory. Both uniform light brown (above) and banded colour forms (below) are shown here.

50

DANGEROUS LAND SNAKES

Dugite

The Dugite (*Pseudonaja affinis*) lives in the south-west corner of Western Australia and along its southern coast into South Australia as far east as Ceduna. It is also found on Rottnest Island and islands of the Recherche Archipelago.

The Dugite is primarily a daytime hunter and is most active in the summer months. It is found in a range of different habitats and is similar in many respects to the other brown snakes in terms of life history. The adoption of the 'S'-shaped defensive pose is also characteristic of this species.

VENOM PROPERTIES

At least one death has been recorded for the Dugite since 1980.

A seventy-two-year-old woman was bitten in Spearwood, a suburb of Perth, although she wasn't aware of it at the time. She started to feel sick after about half an hour and died from a cerebral haemorrhage twenty-two hours later.

The Dugite's venom is strongly pro-coagulant (causing thickening and clotting of the blood) and myotoxic (causing muscle damage).

Ingrams Brown Snake

This species, like the Speckled Brown Snake, is found on the black soil plains of the Barkly Tablelands where Mitchell Grass is the dominant ground cover. There is also a population around Kununurra in Western Australia.

Ingrams Brown Snake (*Pseudonaja ingrami*) is considered to be a small mammal hunter and is noted as preying upon Long-haired Rats in similar

Ingrams Brown Snake (*Pseudonaja ingrami*). A species found on the black soil plains of the Barkly Tableland. This individual is quite darkly coloured. They can also be a light tan or mid brown.

style to the Inland Taipan. Due to its limited distribution in relatively unpopulated areas, bites from this species are extremely rare and there have been no records of fatalities since 1980.

51

AUSTRALIA'S MOST DEADLY AND DANGEROUS BEASTS

Mulga Snakes (*Pseudechis australis*) are the most dangerous of the black snake group and can grow to around 2 metres in length.

BLACK SNAKES

The genus *Pseudechis* includes a number of highly venomous species. To distinguish these species from the previous genus (*Pseudonaja* — brown snakes), they are collectively described here as 'black snakes'. However, one member of this group has a common name that often causes much confusion, the King Brown Snake! Many people tend to associate this species with the wrong genus as a result.

In order to clarify the situation, an alternative common name has been used for the species in this book. It is referred to as the Mulga Snake.

Mulga Snake

Mulga Snakes (*Pseudechis australis*) are quite widespread in Australia, being found over most of the continent with the exception of the coastal strip from Perth east to approximately Ceduna, South Australia, and coastal south-eastern Australia.

Colour patterns are somewhat variable, with central Australian specimens normally olive brown while those in Queensland are brown without any

52

DANGEROUS LAND SNAKES

A Mulga Snake from central Australia, with the olive-brown coloration typical of Mulga Snakes from that area.

hint of olive coloration. Snakes in the southern parts of the range also tend to be darker in colour. Many have a strongly crosshatched pattern of darker-edged scales on the upper body surface.

This is the largest of the black snakes and individuals' sizes increases towards the tropics, so that around the northern parts of the Northern Territory and Western Australia these snakes can be up to 3 metres in length and are sometimes mistaken for pythons. All black snakes tend to flatten their necks into a hood when threatened and arch their forebodies, but they never adopt the 'S'-shaped pose that is typical of the brown snakes. Most large individuals will also hiss loudly when harassed.

Mulga Snakes are active both by day and night and take a range of prey, including other snakes. They lay 11–22 eggs per clutch.

VENOM PROPERTIES

Large Mulga Snakes can yield a considerable amount of venom (180 mg on average) and although it is of relatively low toxicity, it is still potentially very dangerous. Bites normally cause massive tissue damage at the bite site and in major muscle groups (the venom is myotoxic). The venom also has strong pro-coagulant properties. A pressure immobilisation bandage is the recommended first aid treatment (see p. 117).

Spotted Black Snake

The Spotted Black Snake (*Pseudechis guttatus*), or Blue-bellied Black as it is sometimes called, has a restricted distribution in south-east Queensland and northern New South Wales; it is most common on the western side of the Great Dividing Range. These snakes may be completely black or have a series of cream speckles or, occasionally, all the scales can have cream markings, making the snake appear quite pale. The underside is grey-blue.

Spotted Black Snake (*Pseudechis guttatus*). The pattern and number of light grey scales is quite variable on these snakes, with some individuals appearing to be mottled grey (above), while others are black (top).

Spotted Black Snakes feed on frogs, small reptiles and mammals and exhibit many of the standard behaviours typical of the black snake group. When threatened, they arch the forebody and flatten and expand the neck. Loud hissing exhalations of air normally accompany this display.

VENOM PROPERTIES

The venom from the Spotted Black Snake is the most toxic of all the black snake venoms, although cases of snakebite are quite rare. The venom may cause local pain at the bite site and is cytotoxic and myotoxic (it causes tissue and muscle damage).

A Spotted Black Snake exhibiting the defensive pose that is typical of the black snake group. Its flattened neck and forebody is held slightly off the ground in this pose.

Colletts Snake

Colletts Black Snake (*Pseudechis colletti*) is one of the most attractive of our Australian snakes. Its appearance may range from black with crimson markings to grey brown with buff or pale orange blotches or faint rings. The snakes may grow up to 1.5 metres in length.

This species has a very restricted distribution, being found only on the black soil plains of the Tambo region in central-western Queensland. It may be active during the day or at night during the summer and is particularly common after summer rains.

Due to its restricted distribution in a relatively unpopulated part of the continent, bites from Colletts Black Snake are quite rare. In fact, bites are probably more common among snake enthusiasts who keep them as pets than as a result of wild encounters.

Colletts Black Snakes (*Pseudechis colletti*), a favourite with snake collectors, are only rarely encountered in a small area in central western Queensland.

VENOM PROPERTIES

The venom is cytotoxic and myotoxic (causing tissue and muscle damage).

DANGEROUS LAND SNAKES

The Red-bellied Black Snake (*Pseudechis porphyriacus*). One of the most common snakes in Australia, often found in close proximity to water.

Red-bellied Black Snake

Red-bellied Black Snakes (*Pseudechis porphyriacus*) are very common snakes in eastern Australia, their distribution extending from the wet tropics in northern Queensland to Victoria and the south-eastern corner of South Australia. The species is normally found close to rivers, swamps, lagoons and other wet environments where they feed predominantly on frogs and other small vertebrates. They may also hunt small fish and eels, and are normally only active during the day, but they may be encountered on warm summer nights.

As with all members of the black snake group, the Red-bellied Black Snake will exhibit a characteristic defensive pose when threatened, with flattened neck and forebody held slightly off the ground.

As their name implies, the belly is suffused with red along the sides fading to a pink along the middle of the ventral surface. The red colour is more evident in southern specimens and less evident in the north, where it may be quite pale. In these northern regions it is possible to mistake the species for the Small-Eyed Snake (*Cryptophis nigrescens*), which has similar colours.

VENOM PROPERTIES

Red-bellied Blacks are not particularly aggressive and have relatively mild venom. No adults are known to have died as a result of bites, but there are records of child fatalities. The venom is pro-coagulant (causing the blood to thicken and clot), neurotoxic (attacking the nervous system) and myotoxic (causing muscle damage).

Pressure immobilisation bandaging is the recommended first aid procedure (see p. 117).

DEATH ADDERS

Death adders are an interesting group of snakes, quite unlike any of the other dangerous snakes. Since the behaviour and venomous properties are relatively similar for all species, they are discussed as a group rather than individually.

Death adders are comparatively short and stout and rely on their ability to camouflage themselves in their environment to avoid detection. Most habitats in which they occur have a ground cover of leaf litter or grass and they conceal themselves so effectively in it that they are rarely seen. Their colour patterns consist of mottled bands of grey and brown. When threatened, however, death adders expand their bodies and brighter colours at the base of the scales suddenly become apparent, making the snake much more obvious.

Feeding is also something of a curiosity. Since these snakes are not particularly well designed to chase prey, they allow the prey to come to them. In fact, they lure prospective prey — small reptiles, mammals, birds and frogs — within range by flicking the specially adapted tip of their tail, which

Common Death Adders (*Acanthophis antarcticus*) can be highly variable in colour. Shown here is the red form.

resembles a worm. The snake lies in a horseshoe-shaped configuration with the tail close to the head and when potential prey is spotted, the snake wriggles its tail tip enticingly. When the prey comes to investigate, the snake strikes with a single lightning-fast hit.

Death adders have suffered a major decline in their populations since European colonisation, possibly due to habitat destruction and the introduction of the Cane Toad (*Bufo marinus*).

VENOM PROPERTIES

Death adder venom is highly and almost exclusively neurotoxic, causing general paralysis that starts with the lips and then progresses to the other facial and head muscles and then to the body. Before the advent of antivenom, bites were considered to be extremely dangerous, but with modern medical techniques deaths should be exceedingly rare.

DANGEROUS LAND SNAKES

Common Death Adder

This is the largest of the death adder species, with some specimens exceeding 1 metre in length. Common Death Adders (*Acanthophis antarcticus*) are found from the Barkly Tablelands in the Northern Territory through Queensland and into New South Wales, the southern portions of South Australia and the southern coastal regions of Western Australia.

Northern Death Adder

Northern Death Adders (*Acanthophis praelongus*) are found in the northern tropics of Queensland, across to the top end of the Northern Territory and the Kimberley area in Western Australia.

Northern Death Adder (*Acanthophis praelongus*) from northern Cape York and the top end of the Northern Territory and adjacent Western Australia.

Desert Death Adder

As its common name implies, the Desert Death Adder (*Acanthophis pyrrhus*) is found through central Australia, across the Great Sandy Desert to the arid parts of the West Australian coast near the Pilbara. It is usually found in rocky habitats with Spinifex grass.

The Desert Death Adder has recently been split into two distinct species, *Acanthophis wellsi* now being the recognised name for snakes from the Pilbara and North-West Cape in Western Australia. Both species are considered together here as they are very similar in terms of colour, size and behaviour.

This species was reported as responsible for one death in 1998 near Broome in Western Australia. A nine-year-old child accidentally trod on a Desert Death Adder but the bite was not immediately recognised. He died in a vehicle on the way to hospital.

AUSTRALIA'S MOST DEADLY AND DANGEROUS BEASTS

TIGER SNAKES

Australian tiger snakes are grouped together in the genus *Notechis*, with two species formally recognised. There are also a number of sub-species recognised for each of the two species. Some of the sub-species represent isolated populations living on offshore islands along the southern Australian coast.

Black Tiger Snake

The Black Tiger Snake (*Notechis ater*) is found in Tasmania, the islands of Bass Strait, some islands and peninsulas around Adelaide (Kangaroo Island, Port Lincoln) and also in the south-west corner of Western Australia. There is an isolated population in the Flinders Ranges in South Australia.

As the name suggests, these snakes are shiny black and may sometimes have slight banding apparent in a paler shade. The underside is grey.

This young Eastern Mainland Tiger Snake (*Notechis scutatus*) is a light colour form with quite distinctive banding. These snakes are primarily frog hunters and local population declines seem to have occurred in some areas as a result of drought and the decimation in recent years of frog populations by chytrid fungus.

They are found in a wide variety of habitats, including forests, woodlands and grasslands where they feed on small mammals, frogs, birds and reptiles. On offshore islands they are often found living in the burrows of nesting seabirds such as shearwaters and penguins, where they feed on the young chicks.

These snakes are not particularly aggressive or fast but when cornered they will arch the forebody and flatten the neck, in much the same way as black snakes. Toxicity tests place this species around numbers 4 to 6 on the Australian list of most venomous. Venom properties are very similar to those of the Eastern Tiger Snake, which is described opposite.

DANGEROUS LAND SNAKES

A dark Eastern Mainland Tiger Snake. Banding is barely visible in this individual as it gathers the first morning rays of sun in a dew-covered pasture on the Southern Tablelands of New South Wales.

Eastern or Mainland Tiger Snake

The Eastern Tiger Snake (*Notechis scutatus*) is found in Victoria and southern New South Wales, extending up the east coast into south-eastern Queensland. It is a species of temperate climates and in the more northerly parts of its range it occurs as a number of isolated populations on the higher and cooler parts of the Great Dividing Range. Like the Black Tiger Snake, there is at least one island population of Eastern Tiger Snake — it lives on nesting seabirds on Carnac Island in Western Australia.

In the more southerly parts of its range, the snakes appear to be less active on hot summer days, preferring cooler weather with plenty of sunshine. These snakes are quite variable in colour and may be almost black with no noticeable banding, to light tan with yellow bands. Food is primarily frogs, but other prey such as small mammals and reptiles will be taken. Their favoured habitats are along watercourses and in low-lying swampy and marshy areas. Tiger snakes are normally active during the day but will also be found hunting on warm, humid nights.

Eastern Tiger Snakes are not particularly aggressive, but behaviour can vary greatly between individuals. When cornered, the snake will arch the front part of the body in a curve with neck flattened and then strike from that position with a partly sideways lunge.

VENOM PROPERTIES

Eastern Tiger Snakes are second only to the brown snake group in terms of the number of deaths that have been attributed to them; at least four between 1981 and 1991. Prior to the development of antivenom, 50 per cent of all bites were fatal.

The venom is strongly neurotoxic (affecting the nervous system), pro-coagulant (causing blood clotting) and myotoxic (causing muscle damage). Initial symptoms of a bite include pain at the bite site, headache, nausea and abdominal pain. Paralysis is also a common feature due to the neurotoxic properties of the venom; it usually starts as numbness around the lips which progressively spreads to other muscles in the head and then the body. It is vitally important to apply a pressure immobilisation bandage (see p. 117) before any of these paralysing symptoms have progressed.

AUSTRALIA'S MOST DEADLY AND DANGEROUS BEASTS

OTHER POTENTIALLY DANGEROUS SPECIES

The Curl Snake (*Denisonia suta*). These are relatively inoffensive snakes, but large specimens should be treated with respect.

Copperhead

Copperheads (*Austrelaps superbus*) are found in Tasmania, Victoria and along the western slopes of the Great Dividing Range in New South Wales. They are a temperate climate species and can be active in the south at quite low temperatures. They have been recorded above the snowline in the Australian Alps. Copperheads are normally active in the day but will also hunt on warm summer nights.

Their preferred habitats are low-lying swamps and marshes where they can find an abundance of frogs, their primary food resource. They have also been known to take small reptiles and other snakes, including their own species.

Although not particularly aggressive, Copperheads will become defensive when cornered and may bite under these circumstances, although often without much commitment. Their venom is quite toxic but they have relatively short fangs, and so gumboots and stout trousers do offer some protection.

VENOM PROPERTIES

The venom is neurotoxic (attacking the nervous system), pro-coagulant (causing blood to clot) and also myotoxic (causing muscle damage). Interestingly, not many human deaths have been recorded as having been caused by this species.

DANGEROUS LAND SNAKES

Curl Snake

Curl Snakes (*Denisonia suta*) are found in the eastern parts of South Australia and inland areas of New South Wales and Queensland, extending across the Barkly Tablelands in the Northern Territory to the Victoria River area. There is an isolated population in central Australia.

These snakes are nocturnal and terrestrial, feeding on lizards and frogs. Larger individuals on the Barkly Tablelands have been observed feeding on road-killed reptiles and other carrion.

When harassed, Curl Snakes will throw themselves into a series of flattened loops (hence their common name) and will strike if the opportunity is presented. Bites from large individuals (30 cm or more in length) should be regarded as serious and a standard pressure immobilisation bandage used (see p. 117) before seeking hospital treatment.

De Vis' Banded Snake

These small snakes are quite heavily built and are occasionally mistaken for death adders. De Vis' Banded Snake (*Denisonia devisii*) is found in inland southern Queensland and northern New South Wales, where it occurs in association with heavy textured soils in low-lying areas. The snakes' primary food is frogs and they are most commonly seen at night after summer storms.

When provoked, these snakes react in much the same way as the Curl Snake, thrashing around and arching their bodies into a series of flattened curves. They will bite if you get too close. A bite from a large specimen may need medical attention.

Eastern Small-eyed Snake

Eastern Small-eyed Snakes (*Cryptophis nigrescens*) are nocturnal and rarely encountered. They are found from eastern Victoria along the coast through New South Wales and into the southern and central Queensland coasts. A geographically separate population occurs in the wet tropics around Cairns and the Atherton Tablelands.

The average size of these snakes in the southern part of their range is approximately 40–60 cm, but individuals in the tropical north may be up to 1.2 metres in length. Their shiny black dorsal surface and pinkish underside makes them look very much like Red-bellied Black Snakes. They may

De Vis' Banded Snakes (*Denisonia devisii*) are found in central southern Queensland and adjacent areas of New South Wales. Many people mistake them for Death Adders due to their short, squat build.

Small-eyed Snakes (*Cryptophis nigrescens*) are shiny black with a pink underside and may be mistaken for the Red-bellied Black. They have a very different face however, with relatively smaller eyes.

be distinguished by their small eyes (relative to the size of the head) and uniformly pink underside. On Red-bellied Blacks, the red underside is most vivid along the lateral edges and paler in the middle of the belly.

The Eastern Small-eyed Snake prefers moist environments in forests and woodlands, with a preference for rocky outcrops. It feeds predominantly on small lizards.

VENOM PROPERTIES

These snakes may become defensive when provoked and bite if approached too closely. Larger specimens should be regarded as dangerous and their venom is noted as being strongly myotoxic (damaging muscle tissue). One death has been attributed to the species.

Rough-scaled Snake

Rough-scaled Snakes (*Tropidechis carinatus*) look superficially like Tiger Snakes, since they are often prominently banded. They have a limited distribution in south-east Queensland and eastern New South Wales, where they occur in association with dense vegetation along creeks and rainforest edges. A separate population occurs in the wet tropics near Cairns in northern Queensland in similar habitats.

Rough-scaled Snakes feed mainly on frogs but will also take a range of other prey including small mammals, reptiles and birds. They are very active hunters by both day and night and are normally ground dwelling but will climb low vegetation when foraging or basking.

These snakes make good use of their habitat and very quickly disappear into vegetation when approached. If they are cornered, however, they will become very defensive and bite vigorously and often.

VENOM PROPERTIES

The Rough-scaled Snake venom is similar in many respects to that of the Tiger Snakes: it is strongly neurotoxic (attacking the nervous system), but with myotoxic (muscle damaging) and anti-coagulant (thinning the blood) properties as well.

Rough-scaled Snakes (*Tropidechis carinatus*) are found near dense vegetation along watercourses and may often be found sunning themselves on low shrubs, such as the snake is doing in this picture. Their venom has very similar properties to that of the Tiger Snake.

DANGEROUS LAND SNAKES

Stephens Banded Snake

The Stephens banded Snake (*Hoplocephalus stephensii*), lives in northern coastal New South Wales and south-east Queensland. It is quite adept at tree climbing and is most commonly found under bark and in tree hollows. Food consists of a wide range of small mammals, birds, frogs and reptiles.

Stephens Banded Snake (*Hoplocephalus stephensii*).

VENOM PROPERTIES

The venom of these snakes is reported to be highly pro-coagulant (causing the blood to thicken and clot) and neurotoxic (attacking the nervous system). Bites from large specimens should be regarded as potentially serious and medical help sought after applying a pressure immobilisation bandage (see p. 117).

Pale-headed Snake

Pale-headed Snakes (*Hoplocephalus bitorquatus*) are found in coastal southern Queensland and northern New South Wales through a wide range of environments where rainforests, eucalypt forests and woodlands are the dominant vegetation. A separate population occurs in northern Queensland. This species is nocturnal and will forage on the ground or in trees, where it hunts for small reptiles.

Pale-headed Snake (*Hoplocephalus bitorquatus*).

VENOM PROPERTIES

The venom is considered to be neurotoxic (attacking the nervous system) and bites should be regarded as potentially serious.

Broad-headed Snake

The Broad-headed Snake (*Hoplocephalus bungaroides*) has a restricted distribution in the immediate vicinity of Sydney where it lives in close association with sandstone outcrops and ridges. It is a most attractively marked snake and, over the past twenty years or so, has undergone a major decline in numbers due to collecting by snake enthusiasts and the removal of sandstone slabs for gardens and amenity landscaping.

Broad-headed Snakes (*Hoplocephalus bungaroides*) are found in the Sydney sandstone country and live in close association with sandstone outcrops and ridges.

As its name implies, the head is broad and triangular. Both the head and body have an almost black ground colour and are dotted with rows of bright yellow spots. These snakes are likely to be encountered by people who are exploring sandstone areas and may risk a bite from one of these snakes if they lift sandstone slabs without due care. The snake will strike readily when agitated.

VENOM PROPERTIES

The venom is neurotoxic (attacking the nervous system) and bites should be treated as potentially serious.

AUSTRALIA'S MOST DEADLY AND DANGEROUS BEASTS

SOME HARMLESS SNAKES FOR COMPARISON

Unfortunately, every year many harmless snakes and their relatives are killed in the mistaken belief that they are dangerous. All snakes, whether dangerous or not, serve a special role in nature and none should be killed unless it presents such a serious threat to people in its immediate vicinity that no other options are viable. There is no excuse for killing perfectly harmless snakes. The following pages examine a few of the more common snakes that present no real danger.

It must be remembered that many of these harmless species will act defensively if approached and may lunge out and bite an unwary onlooker. In this situation, the most damage you are likely to sustain, however, is a set of teeth impressions and a little shock!

The Bandy Bandy (*Vermicella annulata*) has been implicated in one case of snake bite, but they are a completely docile animal. Of the hundreds that have been captured in wildlife surveys, none have ever been known to bite. They feed on small blind snakes.

Snakes in the genera *Simoselaps*, *Neelaps* and *Brachyurophis* are small, burrowing snakes with shovel-shaped noses. Their colour patterns bear some resemblance to juvenile brown snakes, but all are perfectly harmless to people and domestic pets.

DANGEROUS LAND SNAKES

Pythons

Most people are familiar with pythons (Family Boidae) and have probably seen various examples in zoos or on television. There are no large native python species in Australia that are capable of inflicting injury or death through constriction. Our species range in size from about 40 cm to 4 metres long, although larger specimens to about 9 metres have been recorded.

Pythons may be patterned with various markings or have uniform coloration. All have a series of heat-receptive pits along the upper lip; these are normally quite apparent to the naked eye and are used by the snakes to home in on warm-blooded prey in the dark. Most pythons are quite placid but there are some that will become very defensive and attempt to bite anyone who comes too close. They have no venom and are therefore regarded as harmless.

Australian pythons come in many lengths and colour forms. They are non-venomous constrictors and quite harmless, although they can deliver a good bite when annoyed. Pythons also lay eggs and stay with them until hatched. This level of parental care is not found in any other group of Australian snakes. Shown here is the Carpet Python (*Morelia spilota*) (above), Stimsons Python (*Antaresia stimsoni*) (middle) and the Water Python (*Liasis fuscus*) (top).

AUSTRALIA'S MOST DEADLY AND DANGEROUS BEASTS

File Snakes

There are two species of these aquatic snakes in Australia. They are both members of the genus *Acrochordus* and are found along the northern Australian coast in rivers and estuaries, where they feed on fish and are sometimes caught by anglers using live bait. These snakes rarely attempt to bite and are completely harmless.

Tree Snakes

There are several common species of tree snakes. The two most often encountered are the Green Tree Snakes (*Dendrelaphis species*) and the Brown Tree Snakes (*Boiga*). Both occur in the northern parts of Western Australia, the Northern Territory and from northern Queensland, down the east coast to the Sydney region.

Green Tree Snakes are common in northern Australia. They are often found around dwellings but are rear-fanged and not considered dangerous.

Green Tree Snakes are very fast-moving, daytime snakes and are often seen in trees and shrubs; if on the ground, they will quickly disappear into the nearest available tree. They may spend the cooler months in tree hollows or caves and on occasion, come into houses. The Common Green Tree Snake (*Dendrelaphis punctulata*) is quite variable in colour but typically has a bronze-green upper surface and yellow underside while the Northern Tree Snake (*Dendrelaphis calligastra*) is brown with a cream underside and a dark line extending from the snout, through the eye and down the side of the head. These snakes are quite harmless.

The Brown Tree Snake or Night Tiger (*Boiga irregularis*) has a rather bulbous head with vertical oriented pupils. It is a slow-moving species, usually encountered on warm summer nights as it hunts for small mammals, reptiles and birds. When disturbed, these snakes will recoil into a figure 'S' pose and strike vigorously, but are rear-fanged and essentially harmless.

Whip Snakes

Snakes of the genus *Demansia* are slender, fast-moving terrestrial species. There are eight currently recognised species and one or more of them occur in all Australian States except Tasmania. They range in length from 50 centimetres to 1.5 metres and vary in colour from almost black to pale grey and brown. Most species have white facial markings. Large specimens of the Greater Black Whip Snake (*Demansia papuensis*), which occurs in the northern parts of Western Australia, the Northern Territory and Queensland, are regarded as dangerous by some authorities. All other species are only mildly venomous.

Yellow-faced Whip Snakes are not regarded as dangerous but a large specimen may be able to inflict a bite sufficient to cause considerable discomfort. As with any bite, if generalised symptoms start to appear, medical attention should be sought.

DANGEROUS BIRDS

The first impression that may come to mind is that of a scene from that famous movie *The Birds*, directed by Alfred Hitchcock, but bird attacks are far removed from the realms of science fiction in Australia. Research has shown that around 90 per cent of men and 72 per cent of women have been attacked by magpies at some time in their lives. People are also occasionally attacked by Masked Lapwings or Spur-winged Plovers and, in the tropical north, Cassowaries also feature on the list of badly behaved birds.

Cassowaries (*Casuarius casuarius*) may approach you in areas where they have been fed, but there is no real way of knowing the bird's intentions! Don't risk it! By feeding these birds, you encourage this behaviour and directly contribute to the problem.

Southern Cassowary

Southern Cassowaries (*Casuarius casuarius*) can grow to approximately 2 metres tall and are denizens of our tropical rainforests in far north Queensland. They are flightless birds that forage on the rainforest floor for fallen fruit and vegetable matter.

The wholesale clearing of rainforests has greatly reduced available habitat for the Cassowary and some populations now live in tracts of forest that are in close proximity to urban areas. A good example of such an area is Mission Beach. In such areas interactions with people are not uncommon and the practice of feeding Cassowaries to encourage them for tourism (and also in the belief that they need to be fed to survive) has broken down the birds' natural fear of people, creating a climate where confrontations are frequent. There appear to be several motivators for attack:
- First, a bird may approach people looking for food and then lash out if no food is forthcoming.
- Second, aggression may be territorial, where a bird attacks an intruder who enters its domain. Attacks may also be launched against cars and windows where, presumably, the bird has seen its reflection and interprets it as a rival bird. The colour blue is often associated with attacks on cars and this is possibly a colour that elicits a strong territorial response. It is a dominant colour on Cassowary males' bare neck and lower head regions.
- Third, an attack may occur where a bird is defending its nest or young.

Cassowaries will move quickly towards their intended victim and kick in a downward motion. Their claws are large and can inflict considerable damage. These birds may also peck, crash into or head-butt the victim. In cases where the victim has fallen or is crouching on the ground, the attacking bird may jump onto the victim. Serious injuries may result in these situations and fatalities have been recorded. The last fatality was reported in 1926, so obviously they are quite infrequent. Confrontations normally result in lacerations, but more severe attacks can result in broken bones.

Most Cassowary encounters of the unpleasant kind have occurred when people have been walking along rainforest tracks. There is no evidence that jogging or running is more likely than walking to precipitate an attack.

Avoiding attacks

Cassowaries are wild birds and need to be allowed to live as such in their natural rainforest habitat. Never feed or approach them as you might a tame bird. Be aware that Cassowaries may aggressively defend their nest or chicks and move away from birds that you encounter in the rainforest.

If an aggressive Cassowary confronts you, either stand still or move slowly away while facing the bird. If you have a backpack or other similar padded bag, place it between yourself and the Cassowary. Don't attempt to swing at the bird with a stick or do anything that might be interpreted as aggressive.

After such an event, it is worthwhile reporting the incident to the local Queensland Parks and Wildlife Office. This department keeps records of attacks and may consider moving birds that have a history of frequent attacks against people.

Australian Magpie

From August to November each year our Australian Magpies (*Gymnorhina tibicen*) breed, and it's that time of year when male magpies become protective of their nests and young. Darryl Jones from Griffith University has studied this behaviour recently and has made a number of interesting observations which are recounted below; another source of information is the Injury Surveillance Information System (ISIS), which collates hospital data on magpie attacks as part of its operations. Magpies are found throughout Australia with the exception of the Darwin region and parts of arid Western Australia.

Magpies attack people when they perceive them to be a threat to their eggs or young. Some magpies perceive almost everyone as a threat while others don't recognise anyone as threatening and don't attack at all. Most, however, fall somewhere in between and show varying degrees of selectivity in targeting people. The ISIS database, with over 700,000 records, shows that about 47 per cent of

DANGEROUS BIRDS

attacks target cyclists and 17 per cent target pedestrians. Darryl Jones believes that magpies are more prone to attack when their stress levels rise, and that the faster motion of a cyclist and the associated noise of the bike may tip many magpies over the edge and precipitate an attack.

Victims are more often male rather than female and the majority fall within the age group 10–30 years. Once again this may be due to the fact that men of this age are more likely to ride bicycles in a manner that expends more physical energy, thus being perceived by the magpie to be more associated with threatening behaviour.

Many magpies that attack people have had experiences at some time that have led them to associate people with predation. For example, they may have had children climb up to their nests or their young picked up and taken into care by a well-meaning animal lover. Adult magpies would construe any of these actions as an attack on their nest or young; they retain this memory and then, in future breeding cycles, attack people who resemble those who were involved in the original incident. Thus they tend to target specific individuals and attack them over and over. They also have an excellent sense of time and can adapt readily to the regular traverses of the postman's bike or children going to and from school.

Magpies are able to recognise individual people and it may be possible for someone to alter the attacking behaviour of a magpie if they feed it and act in a non-aggressive way. This may change the magpie's perceptions of that person but it will not reduce the bird's tendency to attack other people!

Magpies (*Gymnorhina tibicen*) look harmless enough, but be careful during the breeding season. About 20 per cent of attacks are directed at the eyes, so wear some protection if you anticipate having to run the magpie gauntlet in your local park or garden.

Avoiding attacks

Magpies will attack people when they perceive that the risk of retaliation from the victim is low. Thus, they tend to attack from behind rather than in front, and watching a magpie will often cause it to defer its attack. A clever way to take advantage of this behaviour is to wear a cap with false eyes on the back. This is effective for pedestrians but putting eyes on the back of a helmet does not seem to work for cyclists! Placing flexible plastic flagpoles on the front and back of a bicycle does work in some instances as the poles may interfere with the magpie's flight path when attacking.

Remember that magpies will generally be in attack mode for only about six weeks of the year, but if their nests are destroyed this may extend the nesting period and so extend their period of aggressive activity.

About 20 per cent of magpie attacks are directed towards the eyes, so if you anticipate being in a belligerent magpie's territory, protective eyewear is highly recommended.

Masked Lapwing

Masked Lapwings (*Vanellus miles*) or Spur-winged Plovers, as they are often called, are found throughout the eastern half of Australia. They will defend their nests quite vigorously by dive-bombing intruders. The nest is a scrape in the ground, usually in an open grassy area, and is camouflaged and often very hard to find. The spurs on the wings of these birds look very menacing, but luckily incidences of actual contact with people during attacks are quite rare. If you are attacked, simply move away from the area and don't attempt to find the eggs or chicks.

Masked Lapwings (*Vanellus miles*) have very impressive spurs on their wings, which are clearly visible in this photograph of a bird preparing to sit on its eggs. Luckily attacks that make actual contact are very rare.

MAMMALS

*Life can be tough in the bush, as the ribs on this dingo (*Canis lupus dingo*) show. Feeding them, however, is the worst thing that you can do! It encourages aggressive behaviour towards people.*

Dingo

DINGOES (*Canis lupus dingo*) HAVE BEEN KNOWN to attack and injure or kill people since the earliest days of settlement, although the frequency of these events was very low. A pastoralist from the Tara area in southern-central Queensland once told of how, many years ago, locals put cowbells on their toddlers when they were playing outside so they could make sure that they knew where they were, as part of their dingo defence strategy! In this same area around 1900, a person was reported to have been eaten by dingoes, but it was unclear as to whether he had suffered a heart attack or had some other event befall him before the dingoes arrived. It is most likely that the dogs appeared on the scene after he had died. All that was found was his boots with his feet still inside them … There are doubtless plenty of other stories like these, told by old-timers in country parts of Australia.

More recently, dingoes have gained notoriety through the case of newborn baby Azaria Chamberlain, who was taken from a tent in a campsite near Uluru (Ayers Rock) in 1980, and the more recent tragic death of a nine-year-old boy on Fraser Island when he was attacked by dingoes in 2001. Reports suggest there have been no less than 400 incidents of dingo attack in Australia within the last five years. To put the issue into perspective, however, remember that there have been far more deaths and injuries caused by domestic dogs over that same period.

Nearly all contemporary incidents of dingo attack involve dogs that have been fed by people. These animals lose their fear of humans and also start to see them as a source of food. When no food is forthcoming and the dogs are hungry, an attack sometimes ensues.

The answer to the problem seems quite simple: don't feed dingoes or attempt to treat them as if they are domesticated. If you are camping, secure all your food at night and when away from the tent site. Place it in strong containers or lock it back in your vehicle. Dingoes are not a problem when they maintain a healthy fear of people. We need to ensure that we do nothing to change that situation!

Platypus

Platypuses (*Ornithorhynchus anatinus*) are animals that are rarely seen and even less often captured or rescued. If however, you do happen to find yourself in a situation where you need to handle a platypus, beware of the venomous spur on the male's back legs (pictured on opposite page)!

The spurs are situated on the inside of the ankle on each back leg. They are hollow and connected to poison glands in the thighs. The usual mode of deployment is for the platypus to drive its back legs together, sinking the spurs into anything that might

Platypuses (*Ornithorhynchus anatinus*) appear to be cute little creatures, and they are if you look but don't touch. The spurs on the inside of the male's hind legs are connected to venom glands and can do considerable damage to an unwary handler.

MAMMALS

be trapped between them. The function of the spurs in normal, everyday pond life is still open to debate. It is thought that they might be used in disputes with other male platypuses or be used to subdue large prey such as frogs.

Most records of human envenomation involve the hands or wrists. The pain can be excruciating, but no life-threatening symptoms have been recorded. The limb may swell and pain may persist for several days. Treat with a cold compress (see p. 116) or pain-relief drugs after consultation with your doctor.

Don't catch or attempt to pick up platypuses. Like all other native wildlife, they are protected and catching them is illegal.

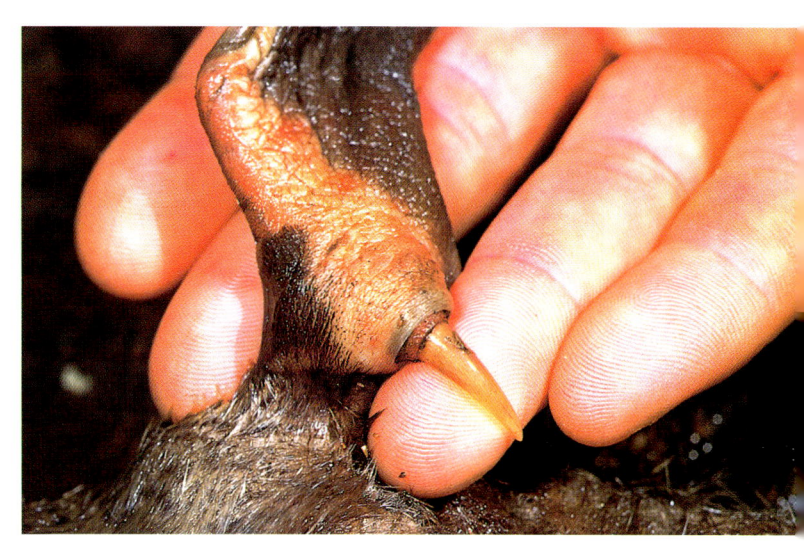

A male platypus's spur is located on the inside of its hind legs. Here it is examined close up.

Water-buffaloes are common throughout the far northern parts of the Northern Territory.

Water-buffalo

Water-buffaloes (*Bubalus bubalis*) were introduced to Australia from South East Asia in the early 1800s and spread through the far northern parts of the Northern Territory. They became very common and widespread in swampy, low-lying environments but in recent years have had their numbers drastically reduced by culling and eradication programmes, as part of the Territory's brucellosis and bovine tuberculosis campaigns. They still occur in reason-

73

able numbers in some parts, particularly Arnhem Land, and care should be taken when you venture into a buffalo area.

Water-buffaloes have been known to attack people, sometimes when accidentally encountered at close quarters and at other times when seemingly unprovoked. Most species of animals are considered to have a 'flight or fight' distance — in other words, they are naturally inclined to run from danger when it is more than a certain distance away but if the threat is closer than that distance, the immediate response will be to fight, or in the case of a Water-buffalo, to charge. It seems that with buffaloes, the fight distance may be quite large — there have been cases where buffaloes have charged from over 75 metres away! In addition, buffaloes will aggressively defend their calves from dogs and other perceived threats, including people.

In about 1983, a Water-buffalo near Darwin charged a friend of the author's and the account still vividly comes to mind. He tried initially to outrun the charging animal but had no chance. As the buffalo arrived, he reached around and, grabbing the horns, threw himself sideways to avoid the brunt of the attack. He got to his feet and ran again, but once again had to fend himself off the horns and throw himself sideways as the buffalo tried to run him down. He did this at least one more time and then tripped and fell. He rolled over and over as the buffalo smashed its head into the ground in an effort to pin him. He jumped to his feet and ran towards a tree, realising that this might be his last chance, but the buffalo was too quick and caught him, tearing his leg open and throwing him to the ground. Face-first in the grass and unable to run any more, he considered his last moments and waited for the end, but the seconds ticked by and no buffalo! After about ten minutes, he lifted himself up to see what had happened and saw the animal grazing peacefully with its back to him, about 45 metres away. He got to his feet and hobbled off!

More recently, in 2001, a well-publicised incident occurred near Jabiru, near Kakadu National Park, when a cyclist accidentally ran into a Water-buffalo on a road at night. He was gored and sustained a broken leg. Four years earlier a man had been killed by a buffalo near the same town when he walked into an animal, also at night.

If you are in an area where Water-buffaloes may be found, be aware of the potential danger. Never walk towards a group of grazing animals and make every effort to avoid surprising a buffalo at close range. Walk in the open and make sure that you are obvious in the landscape. If walking at night, ensure you carry a good torch and, once again, be aware of the potential dangers in startling an animal at close range.

CANE TOADS

THE CANE TOAD (*Bufo marinus*) IS A MEMBER OF the Family that contains the 'true toads', as they are referred to. This Family is widespread in many parts of the world, with a host of different species in the Americas, Asia, Europe and Africa. All share the same stout-bodied appearance. The Cane Toad, or Marine Toad, was introduced to Australia from Hawaii in 1935 as a biological control agent to combat cane beetles in the sugarcane fields of northern Queensland. A native of South America, it had been introduced to Hawaii some time earlier, for the exact same purpose. Unfortunately it turned out to be misconceived plan, and the Cane Toad is now recognised as one of our most serious environmental problems.

The Cane Toad is widely distributed in coastal Queensland and within the next few years will spread through most parts of the Northern Territory and be well on its way to Western Australia. Its rate of spread is estimated to be about 30 kilometres per year. An analysis of its environmental requirements clearly shows it has the capacity to spread over much of Australia, with the possible exception of the arid interior and cool highlands of the south east.

These toads are normally 8–15 cm in length, but in newly colonised areas they can grow to over 20 cm and weigh more than 1 kilogram. They normally hunt at night and will feed on almost anything that moves, provided they can overpower it sufficiently to cram it into their mouths! Cane Toads will also eat dog food and similar items. They breed in temporary or permanent still water

Cane Toads (*Bufo marinus*) are nocturnal predators of most small creatures and can exude a white sticky poison from their parotid glands on the side of the neck. Make sure to wash your hands if you ever make contact with these wrinkled, leathery imports. It should be mentioned that processed leather goods made from toads are perfectly safe!

AUSTRALIA'S MOST DEADLY AND DANGEROUS BEASTS

The poison glands on a Cane Toad will not exude quantities of poison unless placed under pressure during an attack. When this happens, small squirts of viscous white poison will shoot out of each pore on the surface of the gland.

and may breed year-round in the tropics, but are more summer breeders in the south.

Cane Toads have two large poison-secreting glands, one on each side, just behind the head. These glands produce a milky viscous poison that is extremely toxic. It is used primarily in self-defence and will kill would-be predators who attempt to eat the toad. There have been numerous instances where snakes have been found dead with a toad still in their mouths. Many other native animals are known to have died from Cane Toad poisoning, including native quolls, juvenile freshwater crocodiles and goannas. Toads will also poison domestic pets and have been known to kill dogs when the dog has lapped water from a bowl in which a Cane Toad has been sitting.

In people, the venom can cause temporary blindness if introduced accidentally into the eyes and can be absorbed into the body through the eyes, mouth and nose or through cuts and abrasions.

Avoiding Cane Toads

Toads are quite resilient to being picked up and handled and normally don't produce quantities of poison when subjected to this treatment. However, the large neck glands contain a series of pores that open to the surface, and each one contains a quantity of white, viscous poison. If these glands are subjected to pressure, such as when a dog or other animal grasps the toad by the head, the pores squirt a small quantity of venom.

The distance the poison squirts depends upon the degree of pressure applied during the attack. The poison usually squirts into the mouth of the attacking animal, and hence the rather sudden death of 'would-be' attackers. In some cases, people have been squirted with poison when they have attempted to kill toads with shovels and other implements.

Poison can be present on the skin of the toad and poisoning can occur when one's hands are placed near the eyes or mouth after handling a toad. The recommended precaution is simply not to touch Cane Toads. If you do, make sure you wash your hands after the event and don't place unwashed hands near your eyes or mouth in the meantime.

You can prevent toads from getting into your pet's water or food by elevating the bowls or by feeding your pets inside.

The toads' eggs, their tadpoles and the adult toads are all poisonous. Under no circumstances are any parts edible!

Poisoning symptoms

According to information prepared by the Australian Museum in Sydney, Cane Toad venom acts primarily on the heart. Cases of severe human poisoning in Australia are rare, but symptoms in dogs include salivation, twitching, shortness of breath, vomiting and paralysis of the rear legs. Deaths occur after heart failure.

FIRST AID

If you happen to get toad venom in your eyes or mouth, wash thoroughly under flowing water. If you start to develop any other more generalised symptoms, seek medical attention.

MARINE JELLYFISH

AUSTRALIA'S MOST DEADLY AND DANGEROUS BEASTS

Jellyfish are semi-transparent animals with simple body forms that float in the ocean's currents. Using rhythmical beats of their bodies they are able to orientate themselves and adjust their depth, but they have limited ability to control where they may be taken in the ocean. There are, of course, exceptions: the box jellyfish (Cubozoa) are good swimmers and can do more than just float about.

Jellyfish range in size from 1 cm to 2 metres across the bell and can have tentacles trailing for up to 5 metres. The majority, however, have bells that are 2–15 cm across. Most species are short-lived, surviving for only a few weeks or, at the very most, a couple of years.

Their life cycle consists of three main stages. Adults are either male or female and they sexually produce eggs which are held within the adult female's body. A larva hatches from each egg and may either develop within the adult's body or adopt a free-swimming mode, in which case it is known as a 'planula'. The planula finally settles on the sea floor and develops into the next stage, called the 'scyphistoma'. The scyphistoma feeds for some time and grows. It then asexually buds off small medusae which grow and develop into adults.

Some groups of jellyfish never have free-swimming adults. In these groups, the scyphistoma buds off young adults which grow and mature, still attached to the sea floor.

Jellyfish eat a wide range of foods. Some consume plankton while others contain an alga that photosynthesises and, in a mutually beneficial or symbiotic relationship with the jellyfish, allows both to co-exist. Other species of jellyfish catch and eat small fish, worms and crustaceans.

While nearly all jellyfish have stinging cells that enable them to capture food, the box jellyfish, or cubozoans, are the ones with the most potent venom and therefore are most dangerous to people.

Previous page: Blue Bottle, Portuguese Man of War (*Physalia physalis*). Their floating body is about 2 to 3.5 cm long.

CUBOZOANS OR BOX JELLYFISH

The cubozoans are an amazing and fascinating group of jellyfish. While we can appreciate the dangers they pose to humans, it is well worth putting this aspect of their biology aside for a moment to consider some other facets of their lives.

As their common name suggests, box jellyfish are cube- or box-shaped and have either a single tentacle or cluster of tentacles attached to each of the lower four corners of the bell. Being semi-transparent in water, they are often difficult to see.

Cubozoans have excellent swimming ability and do more than float about at the mercy of the ocean's currents. By drawing in and expelling water from the bell, they can move at about human walking pace. Not only are they adept swimmers, but they are also quite agile and can change direction and depth very quickly! This explains, at least in part, why box jellyfish are seldom washed up on beaches in the same manner as other jellyfish.

But swimming ability is only part of the story. Cubozoans have remarkably well-developed eyes that are situated towards the lower part of the bell, in structures known as rhopalia. Some of these eyes have lenses, corneas and retinas just like our own eyes, so their ability to see is probably quite good — although just how good is still a matter of debate. These eyes can look in towards the jellyfish's internal structures or out towards their watery ocean world and cubozoans appear to be able to see underwater obstructions such as piers and rocks, as well as other jellyfish. But how do they see when they don't have a brain to process the visual stimuli recorded by the eyes? This is still a mystery!

Combining the abilities of sight and swimming well, the cubozoans are able to take evasive action when confronted with underwater obstructions and have even been seen trying to avoid capture by marine researchers using hand nets! It is quite likely that cubozoans, including our most dangerous species, try to swim away from humans in the water to avoid contact. Unfortunately they are not always successful. We can compound the chances of contact if we run into the water or move around quickly because this may not give the jellyfish time to escape.

Cubozoans feed on fish and worms and other invertebrates. Their tentacles are ringed with numerous rows of stinging cells called nematocysts. Each cell is like a capsule, inside which there is a coiled-up barb. Upon contact with a small fish, the top of the capsule springs open and the barb is fired, penetrating the fish's skin and releasing venom. The captured fish is then retrieved as the tentacle retracts and the food is transferred to the mouth in the centre of the bell. Nematocysts fire and release poison in the same way upon contact with human skin.

The venom of many cubozoan species is exceedingly toxic. The amount injected depends upon the number of nematocysts that fire and thus is directly proportional to the amount of tentacle that makes contact with the victim.

Australian Box Jellyfish

The Australian Box Jellyfish (*Chironex fleckeri*) is possibly the most dangerous creature in the sea, having accounted for about a hundred deaths over the past hundred years in the Indo-Pacific region. These jellyfish are cube-shaped, like other members of the Cubozoa, and have a bell that grows to the size of a soccer ball. The tentacles — up to sixty in total

The Box Jellyfish, *Chironex fleckeri*, considered to be the world's most deadly. It is a species of estuaries and coastal waters in northern Australia.

— are in four groups, arranged on each lower corner of the bell, and may be up to 3 metres long. Large specimens can weigh up to 6 kilograms.

The Australian Box Jellyfish occurs from the Gladstone area of Queensland, north along the coast, across the Northern Territory and along the Western Australian coast as far as Broome.

The life cycle of this jellyfish is not well known, but polyps (scyphistoma) have been found on stones in the estuaries of rivers in northern Australia. They bud off juvenile medusae from September onwards until the onset of the wet season rains, usually in January.

At this time the young jellyfish are washed into coastal onshore waters — they are never found as far from the coast as the Great Barrier Reef. The peak danger period for people in the water is therefore during the summer months and particularly following heavy rain.

Avoiding stings

The box jellyfish season varies from year to year and it is best to check with local authorities before swimming. Always swim at patrolled beaches and swim inside stinger nets if they are provided.

There are about twenty of these nets in northern Queensland, on some of the most popular swimming beaches. They have a mesh size of 25 mm and are designed to stop large box jellyfish but not the smaller species mentioned below or small segments of tentacles of larger species. Nevertheless, they afford a significant degree of protection.

Special protective clothing can also be worn. This consists of a Lycra suit that effectively prevents the tentacles from contacting the skin and the nematocysts from firing.

Also remember that if someone is stung and requires assistance, the tentacles may be in the water nearby and the rescuer may be in significant danger. But since the jellyfish is likely to swim quickly away after the encounter, the chances of getting stung diminish quite quickly over a short time.

Symptoms

The most immediate symptom of a box jellyfish stinging is extreme pain. It may be so severe that the victim has trouble getting out of the water and may lose consciousness. Red skin lesions appear along the lines of tentacle contact; these may be severe enough to cause permanent scarring. In cases where the sting has been severe, the victim will have trouble breathing and speaking. Heart irregularities may follow, as might cardiac arrest.

FIRST AID

The presence of tentacles remaining on the skin is diagnostic for stings from this particular jellyfish, and while many of the nematocysts have fired and penetrated the victim, many also remain unfired in the tentacle material.

Ring '000' or, if you can get assistance, ensure that someone else does this immediately. The presence of the tentacle will confirm the identification of the stinger and the extent of the wound area will indicate the severity of the sting.

Douse the whole area with liberal amounts of vinegar to deactivate these remaining nematocysts, but be sure not to touch the sting until the vinegar has been allowed to soak. Similarly, stop the victim from touching the area before the vinegar treatment.

Once vinegar has been applied and the nematocysts deactivated, it is safe to touch the wound. The tentacles may then be removed. If the sting area covers a limb, a pressure immobilisation bandage may be applied (see p. 117) to slow the spread of the venom. Remember that time is of the essence with this treatment.

It is estimated that 6–8 metres of tentacle is enough to kill a human. If the victim loses consciousness, check for breathing and heart rate and be prepared to start cardiopulmonary resuscitation.

Antivenom has been developed for these stings and can be administered by suitably trained personnel in hospital.

MARINE JELLYFISH

A group of box jellyfish massed near the water's surface.

Irukandji

This jellyfish (*Carukia barnesi*) is only about 2 cm across and has just four tentacles, one situated on each lower corner of the bell.

Irukandji are widely distributed, and stings have been recorded in Australian tropical waters from Bundaberg in Queensland to Broome in Western Australia. They are not restricted to the coast and stings have occurred at the Great Barrier Reef and elsewhere, at a range of water depths. Judging by the frequency of stings, the Irukandji is a common animal, being more numerous in some seasons than in others. The last period of major activity was the summer of 1991–92, when several hundred people were stung and two deaths were attributed to the Irukandji as recently as 2002. Normally, about sixty people are stung each year and receive hospital treatment.

Quite unlike the previous species, the Irukandji produces only mild to moderate pain when it stings. The true effects of the venom become apparent some half an hour later and the symptoms are often referred to as 'irukandji syndrome'.

Symptoms

Approximately thirty minutes after a sting from an Irukandji, the victim may start to show a series of general symptoms including severe back and abdominal pain and pain in the joints, sweating and agitation, nausea and vomiting. Some degree of paralysis may occur and also high blood pressure and numbness. In the most severe cases, cardiac arrest may follow.

Another Cubozan Jellyfish

Chirpsalmus quadrigatus

This is another cubozoan jellyfish that shares the same body form as the larger Box Jellyfish (*Chironex fleckeri*), in that it has up to nine tentacles at each corner of the bell rather than the single tentacles found on the Irukandji. The bell is about 6–9 cm across. This species is widespread in tropical waters and has been responsible for deaths in the Philippines but none has so far been recorded from Australia.

Symptoms

The venom from this jellyfish is very similar to that of the larger Box Jellyfish, but less potent, in fact, about one-tenth as potent as its larger cousin. Stings usually result in severe local pain but systemic reactions are rare.

FIRST AID

Unfortunately, there is little that can be done in the field, since the major effects are generalised rather than confined to the sting area. The use of vinegar to neutralise the stings has been tried but the results have been inconclusive.

If the sting is on a limb, then a pressure immobilisation bandage may be of some use (see p. 117), although it will need to be applied before the poison spreads and generalised symptoms develop.

Hospitalisation is recommended for all stings that develop in the above-described manner, as Irukandji do cause fatalities in Australian waters.

FIRST AID

Vinegar may be used to deactivate any remaining tentacles and a cold compress used to relieve pain (see p. 116). If other symptoms develop, the victim should be taken to hospital.

Jimble

The Jimble (*Carybdea rastoni*) is a small box jellyfish about 2 cm across the bell with four tentacles. It is widespread in tropical waters and has been known to cause local pain when the tentacles are contacted. Pain normally subsides after a few hours but may leave marks where the tentacles contacted the skin — these usually fade with time.

MARINE JELLYFISH

OTHER JELLYFISH AND JELLYFISH-LIKE CREATURES

Most types of jellyfish, not just the cubozoans, have stinging cells that they use to capture food. Some of these are known to inflict painful stings and some species have actually caused human fatalities. Many species can be listed here: most of them have a body form similar to the Lions Mane Jellyfish (*Cyanea capillata*). It has a dome-shaped bell with masses of fine tentacles trailing behind and has been known to inflict painful stings. Normally the stings of these types of jellyfish are of little concern, however, and can be treated with a cold compress (see p. 116) to alleviate local pain.

A Moon Jelly (*Aurelia* sp.). These are very common and can produce itching if they touch the skin.

Lions Mane Jellyfish (*Cyanea capillata*). This is a species of temperate waters and is considered to be the world's largest, with some specimens measuring more than a metre across the bell. They can produce a burning sensation when they touch bare skin.

Portuguese Man of War, Blue Bottle or Physalia

All of the above are common names for one very well known organism, *Physalia physalis*, that is frequently encountered on surf beaches in Australia. I have referred to it as Physalia in the following discussion. These creatures are technically not jellyfish but hydrozoans. They are composed of a number of different polyps that have joined forces to create a single floating creature; each polyp has a specific role which is vital to the overall functioning of the whole assemblage.

The air sac in hydrozoans is known as the pneumatophore; it is a flotation device that keeps the animal on the water surface. The float also allows the Physalia to be driven by the wind. Not all individuals have the same float configuration and so not all go in the same direction relative to the wind. When you see large numbers of these animals washed up on beaches, remember many more headed the other way out of danger!

The gastrozooids are the polyps that eat and digest food, the gonozooids are the reproductive organs and the dactylozooids are the stinging organs that are used to capture food. The dactylozooid tentacles may be up to 50 metres long on very large specimens!

Avoiding stings

Physalia are sometimes driven inshore by strong winds and since they normally occur in groups, where you see one washed up, there may well be more. If you are concerned about the stings, it may be sensible to keep out of the water at these times. This is particularly the case with young children.

FIRST AID

Tentacles can be removed with tweezers and the sting area treated with a cold compress to alleviate pain (see p. 116). Vinegar is not recommended for washing the stung area.

FISH WITH DANGEROUS SPINES

The Butterfly Cod is a highly ornate-looking fish with venomous dorsal spines.

A NUMBER OF SPECIES OF FISH POSSESS VENOMOUS spines for the purpose of self-defence. Some of these fish are relatively immobile, bottom-dwelling species that rely on superb camouflage to hide from predators and potential prey. They are highly unlikely to move away from approaching danger and use their venomous spines as a second line of defence. Other species are more mobile and swim around, displaying bright colours and highly decorative fins to advertise their dangerous potential to would-be attackers. Still others are more nondescript and we may catch these species when fishing and not realise that they have venomous defence mechanisms until it is too late.

In all cases, the venom is comprised of a number of different proteins, and since these compounds can be destroyed when exposed to heat, immersing the affected limb in hot water is a recommended treatment. Please take time to refer to the information provided on p. 114 before applying this treatment.

Stonefish

There are a number of different species of stonefish in the genus *Synanceia*, which are found in tropical waters around the world. They are regarded as the most dangerous of the venomous spine-bearing fish. Fatalities have been recorded, but in many of the documented cases there have been complicating factors, and the part played by the fish's venom has not been clearly established. Some experts doubt the records of deaths attributed to stonefish and, in fact, consider it highly likely that no one has actually died as a direct result of envenomation.

The most commonly encountered species in Australian waters is the Horrid Stonefish (*Synanceia horrida*), which grows to a length of 25 cm although most are 15–20 cm.

Stonefish (*Synanceia* sp.). Possibly the most venomous fish in the sea. Their colours and the adhering algal growth make them very difficult to see.

FISH WITH DANGEROUS SPINES

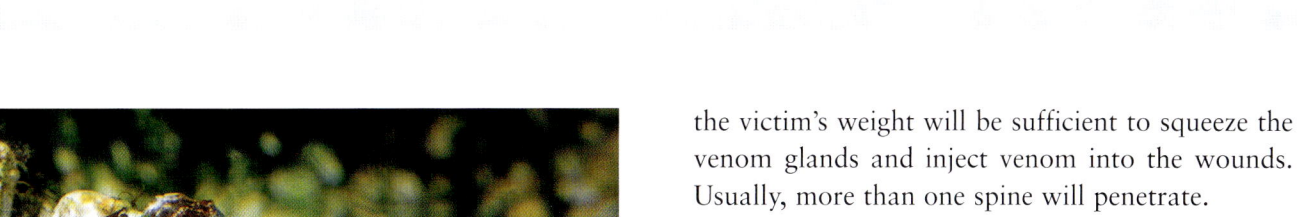

A face that only a mother could love. The stonefish's eyes are to the left and upturned mouth to the right.

Stonefish are sedentary, stout-bodied fish that sit on the sea bottom, usually in rocky or reef habitats. They are perfectly camouflaged and often have algae and other material adhering to their bodies to complete the disguise.

These fish are ambush hunters and wait for smaller fish to pass by before striking with lightning speed. So reliant are they on their camouflage that they will stay quite still, even when there is considerable disturbance nearby. As a second line of defence, they raise their dorsal spines, of which there are thirteen — all of them needle-sharp and connected to dual venom ducts at their base.

Injuries occur when people step on an unseen fish in shallow water. But don't think that you have to be walking in water to get stung! These fish have large, watertight gill covers so they can stay immobile as the tide recedes and sit exposed, swilling trapped water over the gills to breathe, until the water comes back in again.

A foot that comes down on top of the fish will land on the venomous spines, and the pressure from the victim's weight will be sufficient to squeeze the venom glands and inject venom into the wounds. Usually, more than one spine will penetrate.

Avoiding stings

Since these fish remain so well camouflaged, it is unlikely that you will be able to see them and thus avoid stepping on the spines. You might however consider placing your feet only where you can see a clean substrate or a large flat rock when walking through reefs and rocky areas.

Wearing sandshoes or diver's boots is also highly recommended, but don't believe that these provide complete protection. The needle-like spines can penetrate them.

Symptoms

Stings are extremely painful and cause localised tissue damage. The pain may spread up the affected limb and sensitise the lymph nodes, with symptoms peaking about 60–90 minutes after the sting.

Pain may persist for several days but usually subsides after 10–13 hours.

FIRST AID

Immersion of the affected limb in hot water may provide some relief. Make sure that the water does not exceed 45° Celsius, or scalding may compound the problem.

Non-prescription painkillers may be taken. If the victim loses consciousness — this is a very rare occurrence — monitor breathing and heart rate and be prepared to give cardiopulmonary resuscitation (CPR).

Obviously, if a victim's condition deteriorates to this extent, an ambulance should be called. Anti-venom is available for stonefish stings.

Never apply a pressure immobilisation bandage as this may increase local tissue damage.

Scorpionfish and Bullrouts

These fish share a similar sedentary lifestyle with the Stonefish and are well-camouflaged ambush hunters that lie in wait on the sea bottom. Scorpionfish are exclusively marine species while the Australian Bullrout (*Notesthes robusta*) is found in both marine and freshwater environments. Bullrouts are distributed along much of Australia's east coast. In freshwater rivers, they can be found sitting in wait for passing prey under the overhangs of banks. Marine scorpionfish are found in most coastal waters around Australia.

Symptoms

Stings from these fish are similar in many respects to those of Stonefish, although normally much less severe. They result in intense local pain and swelling which subsides after a few hours, or sometimes several days.

FIRST AID

Treatment may include immersion of the affected limb in hot water (not above 45° Celsius) and non-prescription pain killers may be taken. If more general symptoms start to develop, hospital treatment should be sought. As with stonefish, pressure immobilisation bandages are not recommended.

Scorpionfish tend to be bottom dwelling and are sometimes caught by anglers. There are many different species in Australian waters.

FISH WITH DANGEROUS SPINES

Lionfish, Firefish and Butterfly Cod

These are highly ornate-looking fish of the Family Scorpaenidae and are encountered around reefs, particularly in caves and under reef overhangs. For the most part, they are rather slow swimming. Adult fish are predatory and will move in against the reef, using their large fins to trap a small fish against the coral before making a quick lunge. Prey is swallowed whole.

If approached too closely underwater, these fish will erect their dorsal spines and assume a head-down posture. This is a defensive stance which some divers fail to recognise, and the fish may attack if the warning is not heeded. The fish will move forward and ram its spines into a hand or arm.

These fish are also popular in aquariums and many people get stung when cleaning out their aquariums. The mode of attack is the same as that used against divers.

The spines are relatively long and slender in comparison to the stonefish and bullrouts, and their venom is a little less potent, although capable of causing intense pain. The effects usually wear off after 6–12 hours. Treatment is the same as that described for the Scorpionfish and Bullrouts (see p. 88).

A species of Butterfly Cod. Although the banding pattern is distinct from the different species on p. 85, both have the same characteristic body form.

Stingrays

Stingrays (Family Dasyatidae), like the sharks, are cartilaginous fish. They have a flattened body form that is adapted for life on the sea floor, where they forage in the sand and gravel for molluscs and other edible morsels. Food is ground between two crushing plates (rather than teeth, as with most fish). Their gills are on the underside of the body and their eyes are placed close together on the top surface. They are found in all Australian continental waters and there are many species, ranging in size from 30 cm to over 2 metres in length.

Stingrays have an elongated tail that is typically slender, with or without rudimentary fins or appendages, and on the tail is one or more stout, barbed, stinging spines enclosed in a sheath. Stingrays are not aggressive towards people and injuries occur when the rays are stepped on, causing them to strike upwards at the offending leg. The barb is serrated and has two venom-carrying grooves; venom is injected into the wound caused by the barbs. Often the wound is deep, jagged and bleeds freely.

Avoiding stings

Stingrays are normally very wary and will scoot across the sea floor away from approaching people. Sometimes, however, if buried in the sand, they may be approached and are very difficult to see. Nevertheless, your best insurance against a sting is to look carefully where you walk, particularly at night, and drag or shuffle your feet along the bottom rather than bringing them down with each step.

There are many different species of stingrays. All have the same body form and all stingrays should be treated with respect.

The barbed lance from a stingray's tail. It's not surprising that these weapons create considerable damage when thrust into a person's leg.

FISH WITH DANGEROUS SPINES

Symptoms

The wounds are often deep and lacerated and extremely painful as a result of physical damage, as well as the venom that may have been injected. There may be pieces of the spine still embedded. The venom can cause a number of reactions, including nausea and vomiting, muscle cramps, abdominal pain and low blood pressure, and can lead to seizures in extreme cases.

FIRST AID

Immerse the limb in hot water, not above 45° Celsius, to reduce pain and deactivate the venom. If possible, remove any foreign matter from the wound (some of this may be too deep and will need to be removed in hospital, where they may ultrasound the area to locate hidden spine fragments). Do not attempt to close the wound. Leave this for the hospital.

A mass of freshwater catfish (*Tandanus* sp.) vying with one another for a free food handout in Lake Argyle, in Western Australia. Both freshwater and saltwater species have venomous barbs in the dorsal and pectoral fins.

Catfish and Flatheads

These fish are commonly caught by anglers and both have sharp spines that are capable of inflicting significant wounds.

Catfish (Families Plotosidae and Ariidae) may be marine or found in freshwater environments. All are to be treated with respect. They have long, often serrated spines at the leading edge of their dorsal fins (along the back) and also the two pectoral fins (near the gills). These are associated with venom glands which are able to infuse the wounds made by the spine quite effectively. Pain is sometimes very severe and there have been deaths recorded for some overseas catfish species. Immersing the wound with hot water (no warmer than 45° Celsius) is recommended.

Flatheads (Family Platycephalidae) have very sharp spines on either side of the head, just forward of the gill covers. Be very careful when handling these fish and when taking hooks out of them. Treatment for a sting is also immersion in hot water. Even without such treatment, the pain usually subsides within a few hours. Cross-infection by bacteria and other agents is always a distinct possibility with these wounds.

Surgeon Fish

Surgeon fish (Family Teuthidae or Acanthuridae) have a sharp, lance-like projection or spine on each side of their body just in front of the tail. The position of the spine is often marked by distinctive coloration. These spines are used in self-defence and possibly during competitive bouts with other fish when feeding. Great care should be taken when handling these fish as they can cause quite severe lacerations, although the spine has no venomous properties. If you are badly lacerated by a surgeon fish, parts of the spine may be left behind in the wound. In these cases, the wound will require medical attention to clean and close up.

Port Jackson Shark

Port Jackson Sharks (*Heterodontus portusjacksoni*) are sometimes caught by fishermen or encountered underwater by divers. They have a venomous spine on each side of the body, just in front of each of the two dorsal fins. A laceration from one of these spines can result in extreme pain. The effects usually subside after a few hours and hot water immersion (at a temperature not more than 45° Celsius) can be used to deactivate the venom.

Surgeon fish have a scalpel-like spine near the base of the tail. In this fish the position of the spine is indicated by a dark line that runs parallel with the fish's body at the base of the tail.

OTHER SPINY MARINE CREATURES

Some species of sea urchin have long, needle-like spines and can inflict painful injuries.

Sea Urchins

Sea urchins are found in all oceans of the world. They are more or less spherical in shape and are covered in spines of varying design. They also have small pincer-like structures called pedicellariae that are used in feeding. The spines are embedded in a muscular tissue that allows them to move in unison and so produce locomotion. They are also used in self-defence.

Some urchin species can be found in the inter-tidal zone at most times of the day, but others move into rock crevices or under coral and are only found away from their protective shelters at night.

Many urchins have rather stout, blunt spines but others have very sharp, hollow spines that are loaded with toxins. Some species, such as the Flower Urchin (*Toxopneustes pileolus*) have extremely venomous pedicellariae. All inflict painful injuries.

In Australia, most potentially dangerous species are to be found on our tropical coral reefs.

Avoiding urchins

Most injuries result when people accidentally (or sometimes deliberately) touch the spines when wandering about on reefs or in inter-tidal rocky areas. You should always wear good strong footwear and never touch any sea urchins.

Symptoms

It is reported that the spines of some sea urchins contain neurotoxic venom. One would expect a victim to suffer a degree of paralysis and shortness of breath, but such symptoms are exceedingly rare. Most stings cause localised symptoms, including pain and swelling at the site of the injury, which subside after a few hours.

Another noted feature of sea urchin injuries is the development of a hard, inflamed lump at the injury site. These are called sarcoidal granulomas and are the body's defences working to dissolve spine fragments that have been left behind in the wound. These swellings may need to be treated by a doctor if they persist for more than a day or so, as they can cause permanent damage.

Another complication can occur if the spines have penetrated close to a joint in the hands or feet. Parts of the spine or associated material can move into the joint space or tendons and cause tenosynovitis. Indications of such an injury include swelling and soreness of the affected joint. This is an injury that can impair joint function and if not treated can cause permanent disablement. It is important, therefore, to have any spines removed if they have lodged near joints and to seek medical attention as quickly as possible.

Crown of Thorns Starfish

The Crown of Thorns Starfish (*Acanthaster planci*) is familiar to all who have visited the Great Barrier Reef. Periodically, they build up to quite large numbers and some authorities believe that they pose a threat to the survival of the reef as they feed on the coral polyps that create the limestone reef structure.

On the reef, they are most commonly found in sheltered waters at reasonable depth, but will move onto more exposed parts of the reef to feed in calm weather. The Crown of Thorns Starfish is covered with stout spines that are used as a defence against predators.

The Crown of Thorns Starfish (*Acanthaster planci*) has a slimy layer of poison coating its spines.

The surface of the spines is coated with a venomous mucus, composed of substances known as saponins. Injury normally occurs when the starfish is being handled or when they are accidentally trodden on.

Avoiding stings

Be aware of the fact that these starfish have venomous spines and don't attempt to handle them. They are usually quite large and easily seen, so if you are snorkelling or diving, keep an eye out for them and give them a wide berth.

Sting symptoms

The spines will cause severe pain. The site of the injury will often go blue and swelling is also a common feature, followed by numbness. The venom may also cause nausea and vomiting.

FIRST AID

Hot water immersion (the temperature should be no hotter than 45° Celsius) is recommended. The wound may also need to be cleaned and any pieces of broken spine removed. If symptoms become severe hospital treatment may be required, but this is usually rare.

CONE SHELLS

A cone shell hunting. The top tube is the breathing siphon. Below that is the long slender proboscis which is extending from the mouth. The proboscis contains the spear-like radular dart, of which there is an unlimited supply, that is fired into its prey at point-blank range. With small fish, the venom is potent enough to cause almost instant death.

Cone shells are found in all oceans. Approximately 500 species of them are recognised by science and about eighty of these are known from Australian waters. Fortunately, only a few are dangerous to people.

All cone shells are predators. Some feed on worms, others on molluscs and still others are specialist fish hunters. Like all shells, they move relatively slowly and this places the fish-hunting species at a particular disadvantage, since their prey is quite mobile. So to hunt fish, cone shells have developed a very potent venom that can paralyse their prey within seconds, preventing escape. It's these fish-feeding species that are most dangerous to humans.

Cone shells are normally active at night, while during the day they hide under the sand, in coral debris or under rocks. When active, they move along on their large muscular foot in a very similar fashion to a common land snail. At the business end there extends a cylindrical siphon that sucks in oxygenated water to allow the shell to breathe. Below the siphon there is a proboscis that extends out of the mouthparts. The proboscis is long and tapered and can probe around all parts of the shell. Hollow, spear-like 'radular teeth' are produced within the mollusc's body in an organ called the radular sac. These 'teeth' are filled with venom and individually picked up by a muscle that withdraws the tooth from the sac and holds it just inside the proboscis, ready for use.

AUSTRALIA'S MOST DEADLY AND DANGEROUS BEASTS

A cone shell being 'milked' for venom. The tail of a stunned fish is held in front of a small, membrane-covered tube. When the cone shell fires, the radula will go through the thin tail of the fish and penetrate the membrane on the tube, depositing its venom.

The shell uses chemical receptors situated near the end of the siphon to sense the whereabouts of its prey. When prey is detected, the shell moves forward and extends its proboscis until contact is made. The tooth is fired into the prey and venom injected. The process may have several more steps involved for mollusc-feeding shells, but for fish-hunting species it's almost all over.

The spear-like radula tooth is connected to the cone shell with a thread and so the fish is now tethered, with the radula tooth holding fast in the fish's skin. The cone shell retracts its proboscis and draws the fish into the mouth. The fish is dead within a split second and is drawn into the cone shell's mouth. The spear-like tooth is only used once, since it normally remains embedded in the prey. The muscle in the proboscis then retracts into the sac and seizes another tooth in preparation for the next victim.

Avoiding cone shell stings

The vast majority of stings occur when people pick up cone shells. Divers have been known to pick up the shells and drop them into their wetsuits for later examination. The shell gravitates down during swimming and the person may be stung on an unmentionable part of the anatomy!

The commonsense approach here is very simple. If you are going into a marine environment, take time to familiarise yourself with the general form of a cone shell (they are reasonably distinctive) and do not pick them up under any circumstances. Some will tell you that it is safe to pick up a cone shell at the large end. It is not! The proboscis can flex around and reach all parts of the shell if it is held for long enough — twenty seconds may be sufficient! These shells can also sting through clothing and net bags with ease.

Symptoms

The initial sting may be hardly felt or may be similar to a wasp sting. Almost immediately, the area around the bite will lose sensation and numbness may spread up the affected limb or to adjoining parts of the body. The victim may show signs of general paralysis and have trouble speaking. Nausea and stomach cramps have also been recorded in some cases.

Only one fatal case of cone shell envenomation has been recorded in the Australian region. This occurred on Hayman Island in 1935 when a 27-year-old man was stung by a Geographical Cone (*Conus geographicus*). This species has accounted worldwide for about twelve deaths up to 1980.

The other species cited as causing death is the Textile Cone (*Conus textile*), but records are inconclusive. Based on venom analysis, there are a number of other species in Australian waters that are considered potentially capable of inflicting a fatal sting, the moral is to leave them all alone.

FIRST AID

Pressure immobilisation bandages (see p. 117) may be used if the sting is on a limb. If on the body, then general pressure applied over the area may be of some use. If the victim loses consciousness, monitor vital signs and be ready to commence expired air resuscitation or cardio-pulmonary resuscitation (CPR). There is no antivenom available yet.

BLUE-RINGED OCTOPUS

Blue-ringed Octopus (*Hapalochlaena lunulata*) are a dull brown to ochre colour when moving about in search of food. The blue markings only become apparent when they are threatened or disturbed.

There are five recognised species of Blue-ringed Octopus in the world and two of them occur in Australian waters. The larger species, *Hapalochlaena lunulata*, grows to about 20 cm in diameter and weighs up to 100 grams. It occurs around our northern coast. The second species is *Hapalochlaena maculosa*. It is smaller, with a diameter of approximately 10 cm, and occurs along the southern coast.

Both species are crab-hunters and are commonly found in inter-tidal pools and around mudflats. Their hunting technique is rather unusual. Once the octopus finds a small prey, it may simply pounce and hang on to it, but if it finds a larger crab it moves slowly up to it and then squirts venom into the water. The venom immobilises the crab and allows the octopus to move in and feed without risk of injury from the claws. This explains, in part, why their venom is so powerful: in effect, it is a concentrate that is diluted in sea water for use. An individual octopus may contain enough venom to kill two dozen people within minutes.

Blue-ringed Octopuses exhibit a range of different colours which can change to suit their environment. They are normally brown or yellowish brown with mottled markings. When harassed, they turn a darker shade and display a series of electric-blue rings across their bodies and tentacles. These colours are quite striking, and most bites have occurred when people have picked up animals that are displaying the blue rings to show friends, not knowing how dangerous they are. The octopus has a small parrot-like beak and it will use this to bite the unsuspecting person. Venom is produced in the large salivary glands and is introduced into the bite. The first documented death in Australian waters occurred near Darwin in 1954.

Avoiding bites

Treat all octopuses with respect and don't attempt to pick them up. This is especially important if you are fossicking about in the inter-tidal zone or on a reef. Never put your hands into crevices where you can't see what they contain. Remember that when first seen, the octopus may not be exhibiting the bright blue rings and so this feature cannot be used reliably for identification.

Symptoms

The venom contains a neurotoxin known as tetrodotoxin. This is the same as the poison found in puffer fish and is extremely potent. The bite is often not noticed but symptoms will start to manifest within five to ten minutes. As with most nerve agents, the lips will start to become numb, followed by a progressive paralysis of the other facial and head muscles. The victim will become weak and will have difficulty in responding, but still be quite awake and aware of what's going on. The pupils may dilate and finally, respiratory failure will occur.

FIRST AID

If the bite is on a limb, a pressure immobilisation bandage (p. 117) may be used to slow the spread of venom. Once the bandage is in place, the victim will need to be taken to hospital.

Since death is caused by respiratory failure, victims can be successfully treated by maintaining ventilation. If in the field, expired air resuscitation (EAR) should be commenced before heart function starts to become impaired due to lack of oxygen, and will need to be maintained until medical help arrives to take over. With this treatment, the victim should completely recover.

SEA SNAKES

Olive Sea Snakes (*Aipysurus laevis*) are often curious and will approach divers underwater to investigate. Sea snakes are often found in tropical waters.

SEA SNAKES BELONG TO THE FAMILIES Hydrophiidae and Laticaudidae. They are all front-fanged snakes and in this respect they are very similar to our elapid (dangerous) land snakes and most probably evolved from common ancestors. There are thirty-five species found in Australian waters but not all are considered to be dangerous.

Like all other reptiles, sea snakes breathe air and so must come to the water surface every so often to gulp a breath; it appears that they can also absorb oxygen dissolved in the water through their pores. Their lung capacity is huge, due to an enlarged right lung that extends for most of the length of the body. They have upward-facing nostrils that can be closed off when submerged to prevent water entering. They also have a salt-excreting gland under the tongue which enables them to remove ingested ocean salt from their systems.

Sea snakes feed primarily on fish although at least one species is known to be a specialist predator of fish eggs. Most species are found in tropical waters but one, the Yellow-bellied Sea Snake (*Pelamus platurus*), occurs as far south as Tasmania.

Sea snakes have nostrils that can be closed when underwater. This feature is apparent on the Olive Sea Snake pictured above.

Spine-bellied Sea Snake (*Lapemis curtus*). This species is known to have caused human fatalities, although not in Australian waters.

All species have laterally compressed, paddle-like tails. Like their land-based cousins, they can be difficult to identify using colour pattern and size alone. Instead, the pattern and arrangement of scales is normally used by reptile experts in determining correct identification.

Most species forage around reefs and along the bottom in shallow seas, but the Yellow-bellied Sea Snake once again provides an exception, being closely associated with floating weed and rafts of surface debris over deeper waters. Sea snakes are most frequently encountered by trawlers working in tropical waters and they are a particularly common by-catch at night. They are also encountered by divers when exploring coral reefs or sunken wrecks. Less commonly, swimmers come across them or they may be found in rock pools and along beaches after periods of heavy weather. Bites are not particularly common, but have occurred under all of these circumstances.

Avoiding bites

Most fishermen who work in tropical waters are well aware of sea snakes and know to exercise caution when handling nets, particularly when working at night. Divers also need to be aware of the potential dangers posed by these snakes and, during an underwater encounter, it is important not to do anything that might be interpreted by the snake as threatening or aggressive. Snakes that are washed up are best left alone and if you are swimming be aware that any close contacts with sea snakes are to be avoided.

Bite symptoms

The sea snakes' venom is similar in potency to that of black snakes but often bites are shallow and fail to inject venom. In cases where envenomation has occurred, the myotoxic properties of the venom break down muscle tissue and release toxic by-products into the bloodstream. This may result in kidney damage. It may also reduce the body's fluid levels, causing shock. The bite is usually not painful and symptoms develop after 3–6 hours, depending on the position of the bite and its proximity to major vessels.

FIRST AID

First aid measures are essentially the same as those used for dangerous land snakes. Since there is only one type of antivenom in use for all sea snake bites, there is no need to swab the bite site to determine the species involved.

A pressure immobilisation bandage (see p. 117) should be applied as quickly as possible after a bite and the victim taken to hospital for observation and treatment if necessary.

SHARKS

A Great White Shark (*Carcharodon carcharias*) patrolling the cool waters off southern Australia. Mammals such as seals, form a major part of the adult's diet. Great White Sharks can grow to a length of 7 metres.

Sharks (Family Carcharhinidae) are a fascinating and highly adapted group of fish with approximately 160 species occurring in Australian waters. They range in size from the plankton-feeding whale sharks at over 12 metres in length to the Pygmy Shark which grows to about 25 cm and looks similar in many respects to the bottom-dwelling Port Jackson Shark. Other species groups include the wobbegongs, whose fringed margins and dappled colour patterns provide them with perfect camouflage on the bottom. They are ambush hunters and snap up unsuspecting fish as they swim by. Still others are fast-moving, superbly adapted hunters of fish and marine mammals.

Sharks are also of economic importance and are caught and sold in fish markets everywhere. Most of us will have eaten them at some time or another under the general name of 'flake'. Sharks taken by the fishing industry are normally the smaller species such as the Dusky and Sandbar Sharks, although small whaler sharks are also caught.

Sharks take a long time to mature and have slow reproductive rates. This predisposes them to population declines from over-fishing and a number of species are now regarded as threatened, including the Great White Shark.

ATTACKS

From 1803 until 2003, data compiled by Taronga Zoo in Sydney indicates that 625 shark attacks were reported in Australian waters, and of these, 187 were fatal. New South Wales leads the Australian States with a total of 72 fatalities recorded, followed closely by Queensland with 70. Even though these figures may seem impressive, Taronga Zoo points out that many more people die each year through mishaps associated with rock fishing (121 deaths in New South Wales from 1968–1991), surfing (37 deaths in New South Wales in the same period) and other water-related recreational activities.

Shark attacks can be divided into three main types, although some experts disagree with such a classification, arguing that all attacks are similar and should be graded on a continuous scale. For the purposes of a descriptive analysis however, the three-category approach does have some merit and is therefore adopted here.

1. The first type of attack is often referred to as 'hit and run'. These attacks usually involve a single bite and the shark then departs the scene and does not return. They happen most frequently in the high-energy environment of the surf where the water is shallow and the victim's — and possibly also the shark's — vision is obscured by white water and churned-up sand.

   Sharks, when feeding in these environments, move along in the deeper water and make quick forays into the shallows to chase fish that are concealed in the white water. In most of these cases, it is considered that attacks on humans are actually cases of mistaken identity.

   Hit and run attacks are rarely fatal, although this obviously depends on the severity of the single bite. The incidence of such attacks has declined with the introduction of shark nets at many major beaches.

2. The second type of attack occurs in deeper water and is often referred to as 'bump and bite', where the shark will swim up and nudge the victim before attacking. In some cases the victim may escape with minor injuries. In one particular example in the USA, a Great White Shark swam up to a diver and bumped him in the chest several times, then grabbed one arm in its mouth and swam a short distance, released its grip and bumped him a bit more. It then grabbed the other arm and swam along again for a moment, released him once more and swam around. As it passed, the diver picked up a bar that he had been using to collect abalone and struck out. The shark swam away, then turned back and charged. As the shark arrived, the diver managed to grab the edge of its mouth and was taken for a short ride! On releasing his hold, he found that he wasn't far from his boat and so he swam up to it. In the end he only suffered minor cuts to both arms!

   In this kind of attack, it would appear that the shark has recognised the person as an odd type of food item and has been intent on investigating rather than attacking. Sometimes these attacks can result in fatalities since humans are relatively soft, frail creatures and a gentle, exploratory bite can disembowel or do other major damage. There also appears to be an increased risk of provoking a more serious follow-up attack if the victim attempts to fight back during an 'exploratory' encounter.

3. Thirdly, there are full-blown attacks where we see the true power of a top-order predator. The shark charges in from some distance, often unseen by the victim, and hits and bites without hesitation. Depending on the size of the shark and the orientation of the person to the shark at the time, a leg or arm may be taken, or the person may be almost completely eaten. These types of attacks are mostly fatal. There is no doubt that in such attacks the shark has identified the person as a food item and has acted accordingly.

Other shark attacks are thought to be territorial in nature, with particular sharks observed consistently attacking people or other objects within a specific locality.

AVOIDING SHARK ATTACKS

When we consider the number of people who enter the waters around Australia every day, it becomes obvious that the incidence of shark attack is exceedingly rare. In fact, more people get struck by lightning each year! If you are contemplating a holiday at the beach, there are some very simple precautions that you can take to minimise your chances of meeting a shark.

- Don't swim at dusk or dawn, or at night.
- Don't swim near schools of small baitfish (we sometimes see good examples of these on television being patrolled by sharks).
- Don't swim if you are bleeding or menstruating.
- Avoid wearing brightly coloured swimwear or jewellery.
- If nets are provided at a beach, swim within the enclosed area! This may seem obvious, but is too often ignored.
- Avoid swimming near deep-water drop-offs.
- Avoid swimming near seal or sea lion colonies.
- Avoid swimming in murky waters.
- Don't swim where fish, animal or human waste enters the water.
- If spear fishing, never trail speared fish or carry them on your person.
- Don't swim in river estuaries, the lower reaches of rivers or in canals.

SHARK REPELLENTS?

The idea of a shark repellent was first seriously investigated by the US Navy during World War II. The work was prompted by the loss of life occurring in sea warfare and the fact that sharks had been implicated in the deaths of many shipwreck survivors. The first product to be developed was a survival jacket with a pocket that contained a chemical compound composed of black dye and copper acetate. It was known as the 'Shark Chaser'. Following the war, the ineffectiveness of this compound was proved through further research and the repellent was discontinued.

In the 1970s the prospects for a chemical deterrent were revived with the discovery that fish of the sole or flounder group produced a substance that appeared to be highly effective in repelling shark attacks. The compound was analysed and given the name pardaxin. Eventually a number of synthetic substances based on pardaxin were developed. These proved to be quite effective in trials and so it seemed that a commercial repellent was not far away. This was not to be, however, since in sea trials it was found that the chemical rapidly dispersed through the water and became diluted to the extent that it was ineffective. In order to maintain reasonable concentrations, the chemical was put into a squirter canister with a long tube and was tested by squirting it directly into a shark's face. This proved to be an effective repellent, but it assumed that in real life the person would see a shark approaching, have time to orientate the squirter and fire it. The repellent idea was shelved.

There are certainly other marine creatures with repellent chemicals that have yet to be tested as shark repellents, so there is still hope.

Since World War II, experiments have also been conducted using electrical devices, acoustical playback systems and even visual signals as shark deterrents. All have had rather dubious and inconclusive results.

Perseverance may have paid off, however. In March 2002, an Adelaide company launched a new product known as Shark Shield. This system produces an electrical current, and the associated electrical field that results appears to repel sharks. The units are quite compact and can be fitted to a person or surfboard. The system has been tested by strapping large chunks of meat on to a surfboard and pouring blood into the water for good measure. On every occasion, the system has successfully repelled inquisitive sharks.

But some shark experts have their reservations: since the electric field extends only about 4–5 metres from the person wearing the unit, it may be unsuccessful if the shark were to barrel in from some distance away. It is thought that the shark's momentum and focus on the target might be enough to carry it through to contact, in spite of the electrical field that it would encounter in the last second or so of its charge. Unfortunately this type of attack is not easy to replicate in an experiment. Still, the units may be useful under certain circumstances and time will tell. They are commercially available at the present time.

Great White Shark or White Pointer

The Great White Shark (*Carcharodon carcharias*) is the largest predatory shark in the ocean and can grow to a length of 7 metres and weigh over 3000 kilograms. Most, however, are much smaller than this.

Newborn sharks are around 1 metre long and feed primarily on fish. They reach sexual maturity at around nine to ten years if they are male, and around fourteen to sixteen years if they are female. Adult sharks shift their dietary preferences somewhat, and although fish are still important they also rely heavily on marine mammals such as seals and sea lions. They will take birds, for example penguins, and may scavenge offal when it is available. Turtles are also occasionally taken.

The teeth are broad and triangular with serrated edges, about 7–8 cm long, and there are about 300 present in the jaws at any one time. Like other predatory sharks, the teeth are arranged in a number of rows. As teeth in the front row are broken off or worn down, they are replaced by teeth from the rows behind, which rotate up into position. Great Whites use only the front row of teeth in the upper jaw, but in the lower jaw two or three rows are designed to come into play. Thus a Great White Shark bites with approximately eighty teeth.

These sharks are widely distributed throughout temperate waters of the world and seem to prefer water with a surface temperature of between 15 and 22°C. They are normally found singly but do occasionally travel in pairs. Congregations of up to ten have been observed feeding on the bodies of dead whales, but such groups are rare and they rapidly disperse when the food runs out.

There is not a lot known about the movements of Great Whites, but radio-tracked individuals have moved up to 1400 kilometres. Others have been known to frequent the same stretches of water for many years. It is thought that some movements may be seasonal and related to the availability of food.

Killer whales are occasionally known to attack Great Whites, but by far their greatest predator is man. These sharks are often accidentally caught by commercial fishing operations and for many years were specifically targeted by game fishing expeditions. They are also regularly caught in shark nets set along many of our swimming beaches. Their low breeding rate and slow development to sexual maturity does not equip them to withstand this level of increased mortality, and populations have declined over the last fifty years. They were recognised by the Australian Commonwealth Government as a threatened species (Status: Vulnerable) in October 1997 and are also regionally protected in the waters of South Africa, Namibia, the Maldives, Florida, California and in the Mediterranean. It has been speculated that Great Whites may live to about a hundred years of age, but under the current circumstances of high mortality, it is unlikely that any would ever live this long.

Great White attacks

Great Whites are typically ambush hunters and will charge their prey from some distance. Surface swimming penguins and seals are often approached from below, where the shark's grey upper surface camouflages it perfectly against the sea floor. They position themselves and then charge up, grasping their prey and leaping out of the water in the process. This type of attack seems prevalent in some parts of the world, but is not common in Australian waters.

There have been numerous documented attacks on humans by Great Whites and they cover the full range of models (as described on p. 102 under 'Attacks'). Many people have experienced the 'bump and bite' attacks and lived to tell the story, while others have been subjected to the full-on charge and been bitten in half with the shark circling back to pick up the remaining parts.

The Great White is the only species of shark that is known to lift its head out of the water. It is possibly a behaviour that allows them to look for seals and other mammals on the water surface.

They are also noted in some parts of the world for attacking small boats. It is thought that they may be responding to the minute electrical fields created by the metal hulls or the boats' engines. In some cases, boats have been attacked when playing a game fish and, in these cases, it is thought that the vibrations and impulses of the hooked fish may have excited the sharks into attacking. Most of these boat attacks happen in South African waters south of Cape Town, for some unknown reason. One particularly frightening incident occurred at Macassar Bay near Cape Town, where a 5 metre Great White attacked a 6 metre boat and wound up jumping into the boat, landing on one of the occupants, breaking his pelvis and rupturing his bladder. The shark then crashed about, chewing on fuel lines and other items within reach until it finally expired!

Tiger Shark

Tiger Sharks (*Galeocerdo cuvier*) are predators of tropical waters. They may grow to 5 metres in length and weigh up to 650 kilograms. Young sharks are strongly dappled with dark grey blotches and irregular stripes down the sides of their bodies. These markings fade with age, but the body stripes may still be apparent on adult sharks, hence their common name. The snout is rounded and blunt in comparison to the torpedo shape of the Great White.

A Tiger Shark (*Galeocerdo cuvier*). These sharks are usually solitary hunters, and are found in tropical waters.

Tiger Sharks are normally solitary hunters, most commonly found in deeper waters adjacent to reefs. They move over extensive areas when hunting and will venture into shallow water when feeding. Tiger Sharks tend to swim with slow and measured propulsive tail strokes but can suddenly turn on a burst of speed when excited by food.

These sharks are true scavengers of the ocean and will eat most items that they can fit into their mouths. Their staple diet consists of fish and a wide variety of marine creatures such as crabs, octopus, turtles, birds and sea snakes. They also indulge in junk food, in a rather literal sense! When captured sharks have been examined, their stomach contents have been found to include items such as cans, bottles, plastic and metal objects including car number plates. One shark even had a small chicken coop inside, complete with heaps of soggy feathers and bones from the unlucky inmates! If all of this junk becomes a little too much, the shark will disgorge it to make room for new acquisitions.

Tiger Sharks are well known as people-eaters and will often circle their victim before moving in for a kill. They have also been known to use the 'bump and bite' method as described above. They are considered by many to be the second most dangerous shark in Australian waters, after the Great White.

Bull Shark, River Whaler or Swan River Whaler

Bull Sharks (*Carcharhinus leucas*) are robust-bodied sharks with short, blunt snouts that are wider than they are long. They have relatively small eyes that are placed further back on the head in comparison to the other dangerous species. Young are approximately 70 cm long at birth and grow to a maximum size of about 3.5 metres. One large specimen captured near Cairns weighed in at 347 kilograms. For their size, they have relatively large jaws and teeth and are regarded as aggressive predators. Males mature at about 2 metres in length while females are slightly larger.

Bull Sharks are found throughout the world in tropical and semi-tropical waters and are also found in many major river systems. In Australia, they are known from the Sydney area north along the coast and over to the west coast as far south as Perth. Bull Sharks are frequently found in shallow water, normally 30 metres or less, although there are records of them to a depth of some 150 metres.

Throughout their Australian range they frequent river systems and move well up into freshwater

SHARKS

environments. They most probably breed in saltwater habitats but in South America some populations are known to breed successfully in fresh water. Young sharks are found in close association with rivers and estuaries, and in the Brisbane River have been observed to associate in small groups.

In addition to the ability to live in fresh water, these sharks can also go to the opposite extreme and live in waters that are hypersaline — areas where ocean waters have evaporated and salt concentrations have thus increased to well above that of the oceans. The mechanism that allows them to tolerate such a wide range of salinities is currently the subject of investigation at the University of Queensland.

A Bull Shark (*Carcharhinus leucas*) near the Great Barrier Reef, Queensland. These sharks frequently move into very shallow water when hunting.

107

Their diets are quite varied and they have been known to take fish, other small sharks (including their own species), seabirds, dolphins, octopuses, crustaceans and dogs. In fresh water such as the Brisbane River, it is thought that they feed primarily on catfish, bony bream and mullet. These sharks can occasionally be seen around Colleges Crossing on the Brisbane River feeding on mullet.

Bull Sharks are considered by many to be the most dangerous shark of all and they have probably caused many more deaths than have been attributed to them. There are accounts in the USA of them attacking people in as little as 60 cm of water. In one such instance the offending shark was grabbed by the tail and hauled up on the beach by people who had gone to the assistance of the victim. In Australia, they are most probably responsible for a number of recent deaths in the canals around the Gold Coast in Queensland and many of the attacks in surf where identification has not been positively made. Bull Sharks are most active in the summer months and all attacks attributed to this species on the Gold Coast have occurred between October and January.

Some experts believe that an increase in the rate of attacks in rivers in 2001–03 was due to the drought being experienced in parts of the eastern seaboard. It was suggested that the lack of freshwater flow in rivers was allowing salt water to encroach further upstream, thus encouraging marine fish and sharks to move up into swimming and recreational areas. In the Brisbane River it is thought that improvements in water quality, due to erosion and pollution control, have created a healthier environment that has allowed fish stocks to recover. This, in turn, has started to attract predators such as Bull Sharks.

The position and size of the Bull Shark's eyes suggest that the species is probably more reliant on smell and waterborne vibrations to detect prey than vision. It is thus well suited to hunting in the murky, turbid environments of our rivers, canals and estuaries.

Attacks on people have included the 'bump and bite' attacks, where the person is nudged a few times and then 'sampled', and there have also been the out-and-out 'charge and eat' type of attacks.

CROCODILES

The Salt Water or Estuarine Crocodile (*Crocodylus porosus*) is the larger of our two Australian species. It is found from the Rockhampton area of Queensland north along the tropical coast and west to about Broome in Western Australia. Interestingly, there is a population on top of the Arnhem Land escarpment in the Northern Territory. It is also found in the Torres Strait, New Guinea, South East Asia and west across the Indian Ocean as far as India.

The second Australian species is the Freshwater or Johnsons Crocodile (*Crocodylus johnsoni*). These crocodiles grow to a maximum length of about 3 metres and have a relatively slender snout,

Crocodiles will bask to increase their body temperature or seek cool sites to reduce heat. This large croc is employing a slightly different technique and has its mouth open to facilitate cooling by evaporative water loss through the thin tissues inside the mouth.

in comparison to the Salt Water Crocodile. They are found in fresh-water billabongs and creeks in northern Australia and are regarded as harmless. There have been instances where people have been bitten on the legs and feet by these crocodiles, but such attacks appear to be cases of mistaken identity. Their diets normally consist of fish, birds, reptiles, crabs and frogs.

AUSTRALIA'S MOST DEADLY AND DANGEROUS BEASTS

Freshwater Crocodiles (*Crocodylus johnsoni*) have a long slender snouts and are not considered to be dangerous to people.

Salt Water Crocodiles are found in estuaries and along major river systems, extending up to 100 kilometres inland into fresh water. They move around during the wet season and may enter large inland billabongs, lagoons and swamps that then become landlocked during the dry season. They are not confined to rivers and estuaries, however, but are adept ocean goers and have been recorded as travelling up to 1000 kilometres along the coast, between estuaries.

Salt Water Crocodile hatchlings are about 30 cm long and may eventually grow to a length of 7 metres, although most large animals encountered are in the 4–5 metre range. Do any really large crocs exist in northern Australia? It is highly likely, although these individuals would have lived through the time when they were hunted extensively for skins. If they survived that period, it was probably because they were highly secretive and avoided humans, a behaviour that they have continued to the present day. There are some who believe that a number of very large crocodiles still live secret lives in remote stretches of water in the north, far from the attention of humans.

Male crocodiles reach sexual maturity when they are about sixteen years old and average 3.3 metres in length. Females reach maturity at about thirteen years of age when they are about 2.3 metres long. The average life span is thought to be about seventy-five years, and a hundred years is considered to be a maximum.

Salt Water Crocodiles build a large nest of aquatic vegetation that rots down and provides the heat necessary to incubate the eggs. The nests are built and eggs laid during the wet season, from November to March. The female defends the nest and will stay with her hatchlings for a few months after they hatch.

Young crocodiles subsist on a diet of small fish, frogs and similar items. As they grow, food

CROCODILES

preferences become a little more diverse and they will take carrion or wallabies and other animals that may come to the water's edge to drink. Large crocodiles have been known to take fully grown water buffaloes. Animals are normally ambushed at the water's edge after the crocodile approaches submerged. The hit is lightning-fast and the teeth and enormous pressure of the jaws are used to clamp the prey rather than to cut. The animal is then dragged into the water and drowned. If any resistance is encountered, the crocodile will roll to take the victim off its feet. Prey is then torn apart and eaten, or the crocodile may lodge it in a drift of timber or other similar site and come back to eat it some days later.

Before the 1970s crocodiles were hunted for the skin trade and also for recreational purposes. As a result, numbers decreased dramatically and the species almost faced extinction. The situation was recognised by authorities in the early 1970s and by

The Salt Water or Estuarine Crocodile (*Crocodylus porosus*) can move silently and unseen. When on the surface, it can minimise the extent of its exposure, since the shape of the head allows just the nostrils and eyes to protrude.

The jaws of the Salt Water or Estuarine Crocodile are used for clamping large prey rather than cutting. Once clamped, large prey is dragged into the water and drowned.

AUSTRALIA'S MOST DEADLY AND DANGEROUS BEASTS

A saltwater crocodile, barely visible on the water's surface.

1974 all crocodile populations in northern Australia were protected by law. Their numbers have steadily recovered and special management programmes are now in place in areas near human population centres, to control numbers and ensure public safety.

For many years, humans have been the crocodiles' greatest predator. Not surprisingly, therefore, crocodiles tend to maintain a healthy respect for people and keep their distance. In places such as Kakadu National Park, ranger staff maintain regular spotlight patrols and note any crocs that look as if they might be potential troublemakers. They have found that catching these animals and tagging them restores the crocodiles' sense of fear, to the extent that they no longer present a hazard to people.

We might assume that crocs take people because they view them as food, but it appears that this is not always the case as some attacks appear to be motivated by territorial behaviour. There have been accounts of canoeists being attacked by crocodiles that exhibit every indication the encounter is more of a territorial defence rather than a food-hunting foray. Females will also defend their nests and so nesting areas are to be avoided during the wet season!

Seven people were taken by crocodiles in the Northern Territory between 1971 and 2003, and there were sixteen deaths in the whole of Australia during the same period. This is a very minute proportion of the total number of people who would have frequented our tropical rivers and estuaries in that time and the chances of being taken by a crocodile are obviously extremely low. People can, however, place themselves in extra jeopardy and increase their chances of attack if they fail to observe some basic precautions.

Avoiding salt water crocodiles

- Signs at waterholes, wharfs and estuaries warning of crocodiles must be taken seriously. They are not there to add to the tropical mystique of the place! This advice can't be emphasised enough. Remember that ultimately you are responsible for your own safety and you need to read all signage that is provided at rivers, billabongs, lagoons and estuaries — ensure that you obey it.
- Most crocodile attacks have occurred when the victim has been swimming, and the greater proportion of these attacks has been in the late afternoon or at night.
- If there are large crocodiles in an area, don't expect that they are going to be visible! They are highly adapted predators that have refined their hunting techniques and behaviour since the time of the dinosaurs and they can keep out of view quite easily.
- Never clean fish near the river's edge.
- Never tether a dog or play with one at the water's edge.
- Remember that large crocodiles can be patient hunters and have a highly developed ability to observe and deduce regular patterns of activity. If you have a dog and it drinks at the water's edge at a certain place and at a certain time each day, it will be at far greater risk than if its patterns of activity are irregular. The same might be said for people, particularly if your activity involves being near the water's edge at night.

QUICK REFERENCE – FIRST AID SUMMARY

DANGEROUS WILDLIFE SPECIES/GROUP	SUMMARY OF IMPORTANT INFORMATION	FIRST AID TREATMENT
Funnel-web Spiders	Treat funnel-web bites differently to other species' bites (identification information is on p. 7–8). If in doubt, treat the bite as if it is a Funnel-web bite.	Apply a pressure immobilisation bandage, then seek urgent medical attention.
Other Spiders		If required, apply a cold compress to alleviate pain. Watch carefully for more general symptoms: headache, nausea, joint pain, breathing difficulty. If these appear, apply a pressure immobilisation bandage and immediately seek medical attention.
Paralysis Ticks	Tick bites are noted not only for the direct effects of the tick toxin, but also as sites for secondary infection.	Remove the tick, taking care not to squeeze the body (which would force more toxin into the bite). Seek medical attention if victim is exhibiting generalised symptoms. A red spreading rash may indicate a secondary infection that will need treatment.
Scorpions	None of our Australian species is known to be particularly dangerous. The sting will be painful but symptoms will subside after a short time.	Apply a cold compress to alleviate pain. Further treatment should not be necessary unless the victim is particularly sensitive to the sting.
Centipedes	Bites can be extremely painful but not life-threatening.	Apply a cold compress to alleviate pain.
Bees, Wasps and Ants	Single stings are normally not a concern unless the victim is allergic, in which case, treat as if it is a multiple sting. Even non-allergic individuals can develop quite severe symptoms from multiple stings. In the case of honey bee stings, remove the sting from the skin by scraping.	*Single sting, not allergic:* Remove sting. Apply a cold compress to alleviate the pain. *Multiple stings:* The victim may develop anaphylactic shock, exhibiting breathing difficulty, swelling in the throat, lapses of consciousness. If this should start to happen, seek immediate medical attention. Otherwise, a cold compress may help alleviate pain at the site of the stings. Do not scratch the stings as this may break the skin surface and allow secondary infections to develop.

AUSTRALIA'S MOST DEADLY AND DANGEROUS BEASTS

DANGEROUS WILDLIFE SPECIES/GROUP	SUMMARY OF IMPORTANT INFORMATION	FIRST AID TREATMENT
Caterpillars	Hairy and stinging caterpillars normally only cause itching and mild pain.	The stinging hairs may be removed with a sticky material. Cold compresses can be used to alleviate pain.
Land Snakes	Unless you are absolutely positive that the snake is non-venomous, treat it as if it is a potentially dangerous bite. Identification of the snake is not important as the hospital can determine this from an assay of poison on the skin surface around the bite. For this reason never wash the bite site. The immediate application of first aid can be critical.	Apply a pressure immobolisation bandage and seek medical treatment as quickly as possible. Never wait for symptoms to develop before doing this.
Platypus	The Platypus is the only Australian mammal that has venom.	A cold compress may be used to alleviate pain.
Box Jellyfish	*Box Jellyfish* stings occur along coastal beaches or around estuaries. The sting area is covered in long lines of red weals where the tentacles have touched and tentacles will be adhering to the site. The severity of the sting is proportional to the amount of tentacle that has made contact.	Ring for emergency medical assistance. Do not allow the victim or anyone else to touch the sting area as this will fire off extra stinging cells (nematocysts). Liberally apply vinegar to the sting area to deactivate the nematocysts. If the sting area is on a limb, apply a pressure immobilisation bandage after removing the deactivated tentacles. Extreme pain is characteristic of box jellyfish stings. If the victim loses consciousness, monitor vital signs and be ready to administer CPR.
Irukandji	*Irukandji* stings can occur in deep or shallow waters including the area of the Great Barrier Reef. These jellyfish are not normally identified at the time of the sting. The initial sting is not overly painful but general symptoms start to occur about 30 minutes after the sting.	If the initial sting has not been particularly painful, watch for any symptoms that may develop about 30 minutes after the sting. If general symptoms such as back and abdominal pain start to manifest, assume an *Irukandji* sting and seek immediate medical attention.
Other Jellyfish	Normally causing mild pain and/or inflammation	A cold compress may be used to alleviate pain.
Fish with dangerous spines	Injuries inflicted by fish with venomous spines are immediately painful and in the case of the Stonefish, can be quite serious, although deaths are rare. Antivenom is available for Stonefish stings.	Check the wound carefully and remove any broken-off pieces of spine. Immerse wound in hot water but ensure the temperature is not above 45°C. This will help break down the venom.

QUICK REFERENCE — FIRST AID SUMMARY

DANGEROUS WILDLIFE SPECIES/GROUP	SUMMARY OF IMPORTANT INFORMATION	FIRST AID TREATMENT
Sea Urchins	A number of species of sea urchins have long, needle-like spines that carry a toxin. Injuries normally result from treading on or handling them.	Remove the broken spines and seek medical attention if any generalised symptoms start to manifest. Beware of spines that have lodged near a finger or toe joint. Have these referred to a doctor as quickly as possible because they can impair joint function and cause permanent injury. Also seek medical attention if the site of the injury develops a hard lump as this may indicate the remains of imbedded spines that can also lead to complications.
Crown of Thorns Starfish	The spines of this starfish are coated in a venomous mucus that is transferred when an injury from the spines breaks the skin.	Hot water immersion is effective. Ensure that the water is no hotter than 45° Celsius. If generalised symptoms occur, medical assistance should be sought.
Cone Shells	Cone shells are highly attractive and may be picked up by fossickers in the inter-tidal zone or on reefs. They are also found by divers. Several species have very potent venom. Never pick them up. They can sting through layers of clothing and thin wetsuits.	Apply a pressure immobilisation bandage if the sting is on a limb and seek medical assistance. Monitor the victim's vital signs if they are losing consciousness and be prepared to apply CPR.
Blue-ringed Octopus	These octopuses are quite small and when first found, may be a brown or mustard colour without any blue circles or markings apparent. The blue markings appear when the animal is harassed. Never touch small octopuses in the inter-tidal zone.	The venom causes paralysis and death is caused by respiratory failure. Pressure immobilisation bandages may be used if the bite is on a limb. Be prepared to start Expired Air Resuscitation (EAR) if the victim stops breathing and continue it until medical help arrives to take over. With this treatment, the person should make a full recovery.
Sea Snakes	Sea snakes are normally quite inoffensive but may bite when grasped or if dredged up in a trawler's nets. There are quite a few species found in Australian waters and the majority are capable of biting. If bitten, identification of the actual species involved in the incident is not critical, since one standard antivenom is used for all.	If bitten on a limb, a pressure immobilisation bandage should be applied. Medical attention should then be sought.

FIRST AID TECHNIQUES

GENERAL PRINCIPLES

Throughout the book, a number of standard terms have been used to describe first aid techniques. We will look more closely at these in the following paragraphs. Firstly though, we will consider some basic rules of first aid.

St John Ambulance recommends following the DRABC action plan when confronted with an accident scene or emergency. This plan is certainly relevant in the case of potentially dangerous bites and stings.

D — Check for Danger. In all cases, if you are faced with a situation that requires that you go to the assistance of a person who has suffered a bite or a sting or has been attacked by an animal, please take the time to check. Make sure that you are not going to place yourself at risk by moving in to help. If you are also bitten or attacked, you are not going to be able to render first aid assistance and both of you will be in jeopardy. In some cases, you may only need to wait a few seconds for the snake to move off, or for the box jellyfish to swim away. Also, be aware of danger to others around you who may also be willing to move in without thinking.

R — Response. In the vast majority of cases, the victim will be quite lucid, even if suffering from some degree of shock. There are some situations however, where the person may lose consciousness quite quickly. For instance, in the case of a serious box jellyfish sting, the victim may lose consciousness before getting out of the water. Check to see if there is a response. Talk to the person and ask him or her 'Can you hear me?' If the person is conscious, you should then start administering basic first aid required for the particular type of bite or sting.

A — Airways. If the person is not conscious, roll them onto their side and check the airways. With a hand on the forehead, tilt the head back and open the mouth. Clear any obstructions.

B — Breathing. Check to see if the person is breathing. Remember that the patient's condition may, in part, be caused by something other than the bite or sting. For example, a person who has been stung by a box jellyfish may also have taken in a quantity of water, or a snake bite victim may have fallen and hit their head. If the victim is not breathing, you will need to start Expired Air Resuscitation (EAR). If there are obvious signs of a bite or sting, you may be able to get extra help from someone to apply a pressure immobilisation bandage or other specific first aid measure as required to treat the injury while you carry out resuscitation.

C — Circulation. Check to see if the person's heart is beating. Feel for a pulse and if necessary, you may need to start Cardiopulmonary Resuscitation (CPR).

Techniques for EAR and CPR need to be applied by trained operators and will not be described in this book. If you feel that you might ever need to apply these first aid techniques — and most people may be in that position at some time — you should enrol with St John Ambulance and do the basic first aid course. It is really a mandatory requirement for anyone who leaves his or her house!

COLD COMPRESS

A cold compress is simply a freeze brick or jumble of ice cubes wrapped in a cloth to prevent direct contact with the skin. A compress can be used to relieve local pain and reduce inflammation where a bite or sting is not likely to cause anything more than a localised reaction.

HOT WATER

This treatment was briefly described in relation to fish with venomous spines. It can also be used on some other marine stings including injuries inflicted by the Crown of Thorns Starfish. The venoms in these animals are composed primarily of proteins and can be denatured if exposed to heat. However, you will note that in all cases the temperature of the water should not exceed 45° Celsius. This is because it is often very difficult for the victim to sense water temperature with a limb that has had its sensory

FIRST AID TECHNIQUES

abilities compromised by the pain and inflammation from a bite or sting. There have been cases where people have put their limbs into water that was far too hot and caused serious damage, sometimes requiring amputation of the limb. So, using the hot water immersion technique is quite effective, but be very careful.

PRESSURE IMMOBILISATION BANDAGE

This technique, first developed by Dr Struan Sutherland, has revolutionised the treatment of life-threatening bites and stings. The technique is used to prevent venom from spreading throughout the body. The bandage constricts the vessels of the lymphatic system, which is the prime means of transport for venom around the body. The limb is also immobilised, so that muscle contractions and movements are minimised. This also slows down the spread of venom. With the pressure immobilisation bandage in place, blood flow is maintained while the overall movement of venom is reduced. Thus the limb is not starved of oxygen and the bandage can be kept in place for a much longer period of time as compared to the old constrictive bandage technique. The latter technique also had the potential to cause serious complications.

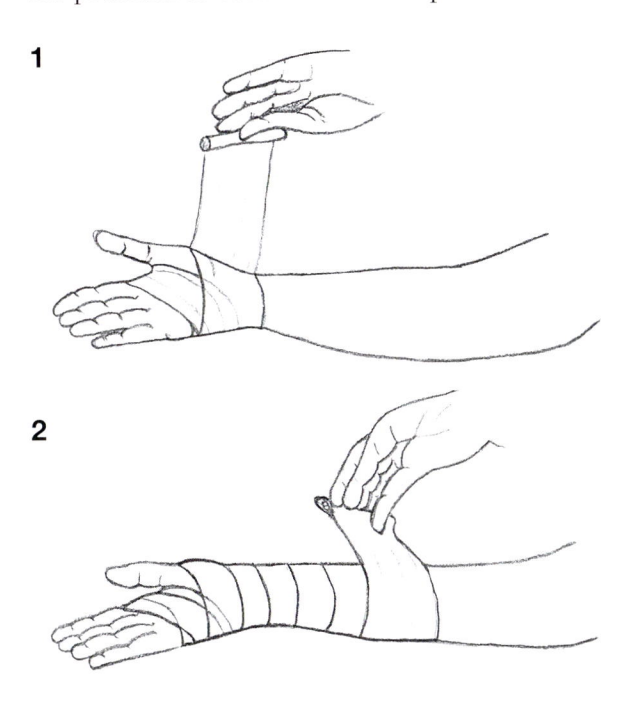

Once a pressure immobilisation bandage is applied, it should be left on until the patient arrives at hospital. The hospital staff will then identify the type of bite. If it was a snake, they will use a snake bite identification kit which assays the properties of the venom around the bite site. They will then ensure that they have the appropriate antivenom and other equipment as might be required and then they will slowly release the bandage and treat the bite. In cases where the bite has not injected any venom, the patient will exhibit no symptoms and thus be given no antivenom.

To apply a pressure immobilisation bandage, you will need a quantity of crepe or conforming bandage about 10–15 cm wide. Start at the end of the limb and wrap the bandage as if treating a sprain. Compress the tissue slightly but don't apply it with so much pressure as to cause blood flow to be restricted. When you have wrapped the complete limb, apply a splint and wrap a second time to secure the splint in place.

Pressure immobilisation bandages are recommended for funnel-web spider bites, dangerous snakes, both land and sea, box jellyfish and cone shell stings.

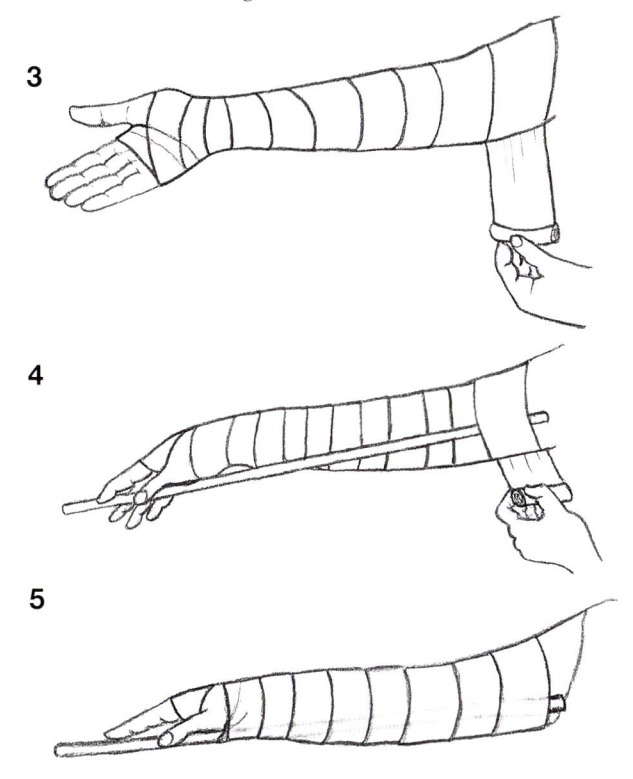

A step-by-step pressure immobilisation bandage technique suitable for arms or legs. An arm bandage may also be placed in a sling thus immobilising the limb and doing away with the splint.

GLOSSARY

Abdomen	The part of an animal where the stomach and digestive organs are located. On a spider, the soft posterior portion of the body.
Anaphylactic (shock)	A severe allergic reaction to venom, characterised by respiratory difficulties, fainting, swelling of the throat and a sudden decline in blood pressure.
Antennae	A pair of elongated sensory organs on the head of an insect or crustacean, etc.
Anti-coagulant	A substance that impairs the ability of blood to form clots (to coagulate).
Antivenom	A substance developed to combat the effects of venom. Usually obtained by repeatedly injecting small amounts of venom into the blood (of an animal) to elicit the production of an antitoxin. The serum is then withdrawn and processed for use to treat people who have received dangerous bites and stings.
Carapace	On spiders: the hard shell covering the cephalothorax.
Cardiopulmonary resuscitation (CPR)	A resuscitation technique that combines expired air resuscitation (EAR) with heart massage in order to maintain blood flow.
Cartilaginous	Pertaining to the possession of a skeleton composed of cartilage, rather than bone.
Cephalothorax	On spiders: the body part that is formed by fusion of the head and thorax (chest) segments. Legs are attached to the underside of this body part.
Chelicerae	The pair of appendages a spider uses to seize and kill its prey. At the end of the chelicerae are the spider's fangs.
Cytotoxic	A substance that destroys body tissue.
Dactylozooids	A polyp that makes up part of the Portuguese Man of War and contains the stinging organs that capture food for the organism.
Denatured	This is when the molecular structure is destroyed, in the case of a protein.
Dorsal fin	The fin on the upper (dorsal) surface of a fish, i.e. along the back.
Elapid	A snake of the family Elapidae. These snakes have fangs at the front of the mouth and all of the Australian dangerous species are included in the group.
Envenomation	The act of having venom introduced into the body by fangs, spines or other injection method.
Expired air resuscitation (EAR)	Resuscitation that involves alternate inflation and deflation of the lungs to maintain oxygen uptake by the body.
Fangs	Modified teeth (of a snake) that are used to inject or infuse venom.

Gastrozooids	A polyp that forms part of the Portuguese Man of War, containing the digestive organs and providing a food processing function for the organism.
General symptoms	Physical symptoms of sickness that involve the whole body or are not restricted to the local area of the bite or sting site. They may include nausea, dizziness, high or low temperatures, difficulty in breathing, etc.
Genus	A level of scientific taxonomic classification. The hierarchy of the classification system includes the Family (usually consisting of a number of genera), Genus (usually including a number of species) and then Species (of which there may be subspecies or varieties).
Gonozooids	A polyp that forms part of the Portuguese Man of War, containing the gonads and providing a reproductive function for the organism.
Histamine	A chemical transmitter that is used to reduce sensitivity to an allergen or toxin.
Hypersaline	Areas where ocean waters have evaporated and salt concentrations have thus increased to well above that of the oceans.
Inflammation	The reaction of the body to an injury, normally including redness, swelling, heat, pain and impaired function.
Life cycle (life history)	The various stages that an organism goes through during its life. For example, many insects start as an egg, progress to a larva, then a pupa and then an adult. These are life cycle stages.
Localised symptoms	Physical symptoms of sickness that are manifest only at the site of the bite or sting. For example, inflammation.
Lyme disease	A disease caused by a bacterium transmitted by tick bites.
Medusae	The adult, free-swimming stage of a jellyfish.
Microenvironment	The environmental conditions (humidity, temperature, etc.) that are created within a small confined area. Usually different to the climatic conditions that exist in the surrounding broader environment.
Myotoxic	A substance that destroys muscle tissue.
Nausea	Feeling of sickness, as if to vomit.
Necrosis	The destruction or death of body tissue in a specific site.
Nematocysts	The stinging cells of a jellyfish that fire a coiled poisonous barb upon contact.
Neurotoxic	A venom or poison that attacks the central nervous system and impedes nervous system functions such as breathing.
Oviparous	Producing ova or eggs, which hatch after being expelled from body.

GLOSSARY

Pathogen	A damaging agent that causes impairment to the human body.
Pectoral fins	Fins on the pectoral girdle (breast) of a fish. These are the fins that are located on each side near the gills.
Pedicellariae	Small, pincer-like structures used by sea urchins in feeding.
Pheromone	A chemical scent that is produced by an organism that acts as a 'messenger' to other members of the same species (or occasionally other species).
Planula	The free-swimming larvae of some jellyfish.
Poison	A substance that impairs or threatens life and that is taken by mouth or otherwise ingested but not injected.
Polyp	An organism with a tubular sac-like body and mouth and tentacles.
Polyvalent antivenom	An antivenom that is designed to combat the effects of a number of different venoms.
Predation	The act of hunting and killing prey animals.
Proboscis	Modified mouthparts of an insect or other animal that are elongated into a tube for feeding.
Pro-coagulant	A substance that causes blood to thicken and clot.
Q fever	An infectious disease characterised by high temperatures and a long period of recovery, caused by the organism *Coxiella burnetti* thought to be transmitted by ticks.
Radula sac	The sac in which the radula teeth are formed.
Radula teeth	In molluscs (snails and their relatives), the hard surfaces in the mouth that are used in grazing vegetative material. This is modified in cone shells to form a spear-like structure that is dipped in poison and thrust into prey to capture it.
Rhopalia	Structures found in the lower part of the cubozoan jellyfish bell, where the eyes are located.
Rickettsial spotted fever	Disease caused by *Rickettsia*, bacteria-like micro-organisms thought to be transmitted by ticks.
Sarcoidal granulomas	The hard inflamed lumps, which form as a local reaction to sea urchin spines. When formed near joints, they can cause debilitating permanent injuries.
Scyphistoma	The name given to jellyfish in the second main stage of its development, the polyp stage.
Sensitised	To elicit a more pronounced adverse reaction to a venom or poison than might normally be expected. People who are sensitised may have had previous contact with the venom or poison and have developed a heightened sensitivity.

GLOSSARY

Siphon	A tube-like structure through which water is drawn.
Species	The most fundamental category used to describe a single kind of plant or animal. Members of the same species can breed together and produce fertile offspring.
Spinnerets	The web-producing appendages at the base of a spider's abdomen.
Tenosynovitis	Inflammation of the tendon sheath.
Thorax	In insects: the body segment that supports the legs.
Toxin	Poison.
Ulceration	An area of skin and underlying tissue that has disintegrated.
Vector	An organism that may carry and transmit a disease.
Venom	A poison that is introduced into the body through a bite or sting (is injected).

INDEX

Africanised Honey Bee 30
ants 34–6, 113
assassin bugs 37
Australian Bullrout 88

Banana Spider 4
Bandy Bandy 64
Barking Spider 12
bees 30–1, 113
Bird-eating Spider *see* Barking Spider
birds 67–70
Black House Spider 15–18
black snakes 52–5
Black Tiger Snake 40, 58
Black Widow 4
Blue Bottle 77, 84
Blue-ringed Octopus 97–8, 115
Box Jellyfish 78–80, 82
Broad-headed Snake 63
brown snakes 47–51
Brown Tree Snake 66
Bull Shark 106–8
Bullrouts 88
Butterfly Cod 85, 89

Cane Toad 75–6
caterpillars 38, 114
catfish 91
centipedes 27–8, 113
Chirpsalmus quadrigatus 82
Coastal Taipan 40, 41, 45–6
Colletts Black Snake 54
Common Death Adder 56, 57
cone shells 95–6, 115
Copperhead 60
crocodiles 109–12
Crown of Thorns Starfish 94, 115
cubozoans 78–9, 82
Curl Snake 60, 61

Daddy Long-legs Spiders 19–20
De Vis' Banded Snake 61
death adders 56–7

Desert Death Adder 57
Dingo 71–2
Dolomedes 2
Dugite 51

Eastern Mouse Spider 12
Eastern Brown Snake 40, 47–8
Eastern Mainland Tiger Snake 40, 58, 59
Eastern Small-eyed Snake 61–2
Estuarine Crocodile 109–12
European Wasp 31, 33

file snakes 66
firefish 89
first aid
    summary 113–15
    techniques 116–17
fish with dangerous spines 85–92, 114
flathead 91
Flower Urchin 93
Freshwater Crocodile 109
funnel-web spiders 2, 4, 5–9

Golden Orb-weaver Spider 19
Great White Shark 101, 104–5
Greater Black Whip Snake 66
Green Tree Snake 66
Greenhead Ant 34

Honey Bee 30–31
Horrid Stonefish 86–7
Huntsman Spiders 20

Ingrams Brown Snake 51
Inland Taipan 40, 46
Irukandji 82

jellyfish 77–84, 114
Jimble 82

Katipo Spider 4

Lionfish 89
Lions Mane Jellyfish 83

INDEX

Magpie 69–70
Malmignatte Spider 4
mammals 71–4, 114
marine jellyfish 77–84, 114
Masked Lapwing 70
Mouse Spider 11
Mulga Snake 52–3
mygalomorph suborder 4, 11–12

Night Stinger Spider 4
Northern Death Adder 57
Northern Tree Funnel-web 5

octopus *see* Blue-ringed Octopus
Olive Sea Snake 99
orb-weavers 19

Pale-headed Snake 63
Paper Wasp 31, 33
paralysis ticks 21–4, 113
Platypus 72–3, 114
Port Jackson Shark 92
Portuguese Man of War 78, 84
pythons 65

Red-back Spider 1, 4, 13–14
Red-bellied Black Snake 55
Red Fire Ant 35, 36
Rough-scaled Snake 62

Salt Water Crocodile 109–12
Scorpionfish 88
scorpions 25–6, 113
sea snakes 99–100, 115
sea urchins 93–4, 115
Selenocosmia stirlingi 2
sharks 101–8
snakebite 44
snakes, land 39–66
    avoiding unpleasant encounters 42–3
    black snakes 52–5
    brown snakes 47–51
    death adders 56–7
    first aid 44, 114

harmless snakes 64–6
how we compare with the rest of the world 41–2
identification 40
inter-breeding 44
most 'dangerous' 40–1
'nests' 43
taipans 45–6
tiger snakes 58–9
Southern Cassowary 67–8
Speckled Brown Snake 50
spiders 1–20, 113
    anatomy 3
    first aid 113
    funnel-web 2, 4, 5–9
    harmless spiders 19–20
    identification 2
Spine-bellied Sea Snake 100
Spotted Black Snake 53–4
Stephens Banded Snake 63
stingrays 90–1
stonefish 86–7
surgeon fish 92
Sydney Funnel-web 2, 4, 7

taipans 45–6
ticks, paralysis 21–4, 113
Tiger Sharks 105–6
tiger snakes 58–9
tree snakes 66

wasps 31–3, 113
Water-buffalo 73–4
Western Brown Snake 49
whip snakes 66
White-tailed Spider 15–18
Wolf Spider 15–18

Yellow-bellied Sea Snake 99